Green Energy and Technology

For further volumes:
http://www.springer.com/series/8059

Mark J. Kaiser · Brian F. Snyder

Offshore Wind Energy Cost Modeling

Installation and Decommissioning

Springer

Mark J. Kaiser
Center for Energy Studies
Louisiana State University
Energy, Coast and Environment Building
Baton Rouge, LA 70803
USA

Brian F. Snyder
Center for Energy Studies
Louisiana State University
Energy, Coast and Environment Building
Baton Rouge, LA 70803
USA

ISSN 1865-3529
e-ISSN 1865-3537
ISBN 978-1-4471-2487-0
e-ISBN 978-1-4471-2488-7
DOI 10.1007/978-1-4471-2488-7
Springer London Heidelberg New York Dordrecht

British Library Cataloguing in Publication Data
A catalogue record for this book is available from the British Library

Library of Congress Control Number: 2011944771

Printed on acid-free paper

Springer is part of Springer Science+Business Media (www.springer.com)

Preface

In Northern Europe, a significant fraction of wind power growth in the past decade has occurred offshore where winds are typically stronger and more constant; by contrast, in the U.S. there are currently no offshore wind farms operational or under construction. The slow progress in offshore wind development in the U.S. has been attributed to several factors including challenging economics, less government involvement, environmental concerns, lack of public acceptance, poorly capitalized development companies, and the lack of a regulatory system for leasing. By late 2011, the Bureau of Ocean Energy Management, Regulation and Enforcement had created a regulatory system for offshore wind development, several developers had signed power purchase agreements and formed partnerships with larger companies to facilitate financing, and public resistance had subsided in many areas. Construction of several offshore wind farms is expected within the next decade.

The purpose of this book is to develop models of two related costs experienced by offshore wind developers: installation costs and decommissioning costs. Installation costs are the second largest component of capital expenditures for offshore wind after procurement and play an important role in determining the overall profitability of a development. Decommissioning costs, while expected to be a fraction of installation costs, are important to policy makers because they determine the financial assurance regulators will require before allowing development to proceed. Identifying the proper value of this assurance is important; if the assurance is set too low, the government may be left financially responsible for decommissioning operations; if the assurance is set too high, an unnecessary cost for developers is created.

Chapters 1 through 4 provide introductory and background material. In Chap. 1 the current status of offshore wind farms in Europe and the U.S. is discussed and information on generation capacity and capacity growth in Europe are presented. The European market holds more than 95% of global cumulative installed offshore capacity and is the dominant world market, and given the number of installation vessels being delivered to European firms, future large capacity additions are probable. In Chap. 2 the system components of wind farms are defined, and in

Chap. 3, the general stages of wind development are reviewed from a contractual perspective and offshore state and federal leasing regulations are highlighted. Chapter 4 provides a conceptual basis for subsequent chapters whereby the factors that influence project costs, cost estimates, and liability are discussed and comparisons to the European wind market and U.S. oil and gas market are made.

Chapters 5 through 7 provide the technical details used to inform the cost estimation modeling. In Chap. 5, a primer on installation methods is provided and data on installation times for turbines, monopiles and cable projects in Europe are analyzed. This analysis is used to inform estimates of installation costs and decommissioning times and sets a baseline for operational activity. In Chap. 6, the vessels required for installation and decommissioning, their dayrates and required spreads are discussed. No U.S. market currently exists for turbine installation vessels, and so a central question for developers is the expected cost of vessel dayrates. In Chap. 7, we address this question and develop empirical models of installation vessel dayrates and mobilization costs.

In Chap. 8 a reference class approach is used to normalize and quantify the capital costs of offshore wind farms in Europe. We describe how European costs may differ from those experienced in U.S. markets and the factors that impact the differences. The estimates presented provide the context for installation and decommissioning costs.

In Chap. 9 a model of installation cost is developed based on the expected duration of installation activity and the daily vessel costs. The installation costs at three planned U.S. wind farms (Cape Wind, Bluewater Delaware, and Coastal Point Galveston) are estimated and sensitivity analysis performed to identify the variables most responsible for uncertainty and risk.

Chapters 10 through 12 focus on decommissioning. In Chap. 10, the regulations that specify decommissioning bonds, the stages of decommissioning, and expected work flows are outlined. An alternative method for turbine removal which involves felling the turbine like a tree, rather than removing it piece-by-piece with a large elevating vessel, is proposed. The formulation of viable alternative removal methods such as felling illustrates the uncertainty in decommissioning procedures and cost estimation early in the life cycle of development.

In Chap. 11 component weights are estimated for the structural components of wind farms. These weights are used to estimate the scrap value and disposal costs of decommissioning in Chap. 12. Chapter 12 uses models similar to those developed in Chap. 9 to estimate removal and disposal costs. The scrap value of steel in the foundations, towers, and turbines is included and the costs of alternative removal options are modeled. The models are parameterized with available data on proposed U.S. offshore developments.

Acknowledgments

The material in this book was supported in part by a grant from the U.S. Department of Interior under MMS Contract M10PC00065 from the Bureau of Ocean Energy Management, Enforcement and Regulation (BOEMRE). The contents do not reflect the views or policies of the BOEMRE, nor does mention of trade names or commercial products constitute endorsement or recommendation for use.

Contents

Chapter 1
Offshore Wind Development 2011

1.1 Offshore Wind in Europe

European offshore wind farms operational and under construction in 2011 are shown in Fig. 1.1. In total, approximately 6 GW of capacity is operational or under construction, and represents more than 95% of global cumulative installed offshore wind capacity. In Fig. 1.2, the number of farms built by year and the average capacity of new farms is shown and in Fig. 1.3 the cumulative capacity and annual capacity addition is depicted. Data for 2012–2014 are based on developer estimated online dates.

Prior to 2009, less than five projects with a total capacity less than 250 MW came online each year. In 2009, there was a dramatic increase and 11 farms with a total capacity of approximately 1 GW came online. From 2010 through 2014 an average of 2 GW is expected online each year and the total capacity expected to be online by 2014 is 12 GW.[1] Capacity additions in 2014 are likely an artifact of optimistic planning, but growth trends are likely to continue as long as the political will exists to support development and the infrastructure required to support the industry continues to develop.

Table 1.1 shows offshore wind capacity by nation and total consented, under construction, and generating capacity. Roughly half of the total 23.7 GW capacity is expected to be online by 2014 and the other half added before 2020. As of December 2011, the UK had the largest wind generation capacity in Europe with 1,341 MW of generating capacity and 2,239 MW under construction.[2] This represents 45% of European offshore wind capacity and 75% of capacity under construction. Including consented projects, Germany has 8.6 GW of offshore wind

[1] This only includes wind farms that are consented and for which an estimated date online is available.

[2] In 2011, offshore wind is expected to account for 1–2% of UK electricity generation assuming a capacity factor of 35% and total national generation similar to 2005–2010 average.

M. J. Kaiser and B. F. Snyder, *Offshore Wind Energy Cost Modeling*,
Green Energy and Technology, DOI: 10.1007/978-1-4471-2488-7_1,

Fig. 1.1 Wind farms over 100 MW online and under construction—December 2011 *Data* from 4COffshore, industry press

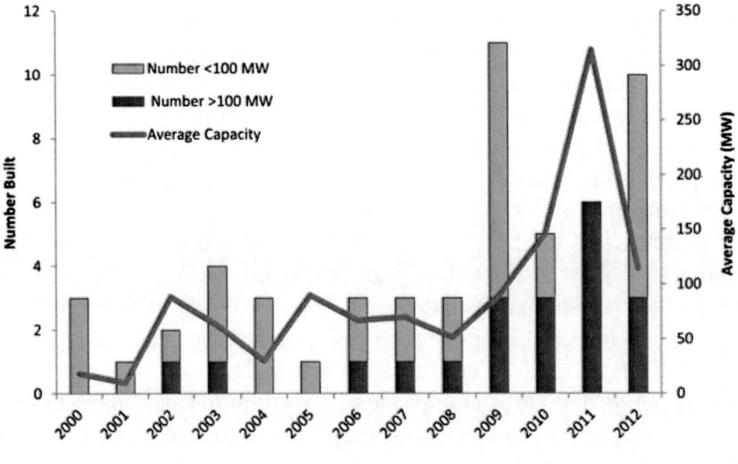

Fig. 1.2 Number of offshore wind farms built per year and average new farm capacity *Data* from 4COffshore, industry press

capacity which, if built, will account for 4–5% of national generation. Thus, in the near to mid-term offshore wind in Europe will be a small but not insignificant contributor to total electrical generation.

Growth in the European offshore market will depend principally on the ability of developers to manage cost and the supply chain to meet demand. Growth across the entire supply chain is required if European nations are to meet national targets. Specifically, newbuilt installation vessels, additional foundation manufacturing capacity, and the development of specialized offshore wind turbines with their own manufacturing supply chain are required [1].

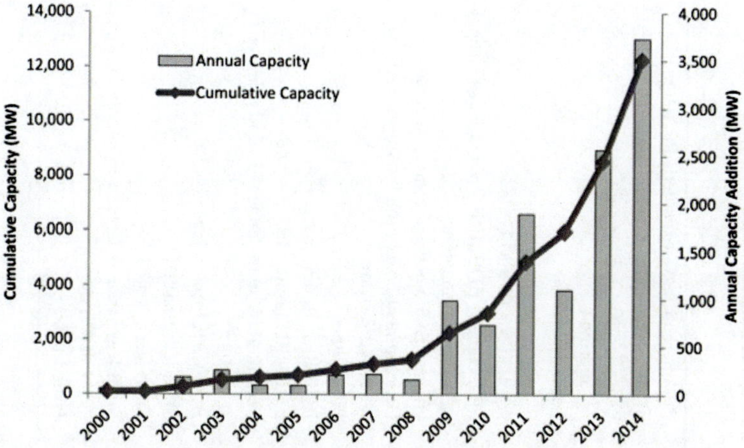

Fig. 1.3 Annual additions and cumulative offshore wind capacity *Data* from 4COffshore, industry press

Table 1.1 Offshore wind capacity by nation—December 2011

Nation	Consented (MW)	Construction (MW)	Operational (MW)	Total (MW)
United Kingdom	1,257	2,239	1,341	4,837
Denmark	436	0	856	1,292
Belgium	529	148	195	872
Netherlands	3,037	0	228	3,265
Sweden	1,531	0	161	1,692
Germany	7,909	600	121	8,630
Finland	768	0	30	798
Ireland	1,100	0	25	1,125
Norway	407	0	2	409
Estonia	700	0	0	700
France	108	0	0	108
Total	17,782	2,987	2,959	23,728

Data from 4COffshore, industry press

1.2 Offshore Wind in the U.S.

Offshore wind development in the U.S. is planned for the Atlantic Coast, Great Lakes, and Gulf of Mexico (Table 1.2; Fig. 1.4) but to date, no turbines have been installed. A number of projects are planned, but it is uncertain how many will eventually be developed and it is unlikely that projects which have not made significant progress by 2012 will be online before 2015. Despite favorable wind speeds, development along the Pacific Coast is unlikely in the short term due to the depth of the near-shore Pacific shelf.

Table 1.2 Proposed U.S. offshore wind farms and development status—December 2011

Developer	Wind park	State	Jurisdiction	Capacity (MW)	Online by 2015*	Status
EMI	Cape Wind	MA	Federal	468	L	Commercial lease executed; PPA signed; awaiting financing
Coastal Point	Galveston Offshore Wind	TX	State	150	L	Lease issued; waiting on Corps of Engineers approval
Coastal Point	Jefferson, Brazoria, Corpus Christi, and Brownsville	TX	State	1,800	U	Lease issued
LINYC	Long Island Offshore Wind Park	NY	Federal	350–700	U	Applied for BOEMRE lease
Bluewater Wind		DE	Federal	200–600	L	PPA signed; BOEMRE limited lease issued; awaiting financing
Southern Company		GA	Federal		U	Offered BOEMRE lease; not executed
Hull Municipal	Hull Offshore Wind	MA	State	12–20	U	Development costs prohibitive; unlikely to move forward
Deepwater Wind	Block Island	RI	State	30	L	PPA signed
Deepwater Wind	Garden State Offshore Energy	NJ	Federal	350	P	BOEMRE limited lease issued
Fisherman's Energy		NJ	Federal	330	P	BOEMRE limited lease issued
Fisherman's Energy	Atlantic City	NJ	State	20	L	Permitted by state; launched wind monitoring buoy
Bluewater Wind		NJ	Federal	350	P	BOEMRE limited lease issued
Baryonyx	Rio Grande	TX	State	1,000–1,200	U	State lease issued
Baryonyx	Mustang Island	TX	State	1,000–1,200	U	State lease issued
GLOW/LEEDCo	Cuyahoga	OH	State	20	L	State lease issued; agreed to supply contract with GE
NYPA		NY	State	120–500	U	Issued RFP

Note * L denotes likely; P denotes possible; U denotes unlikely

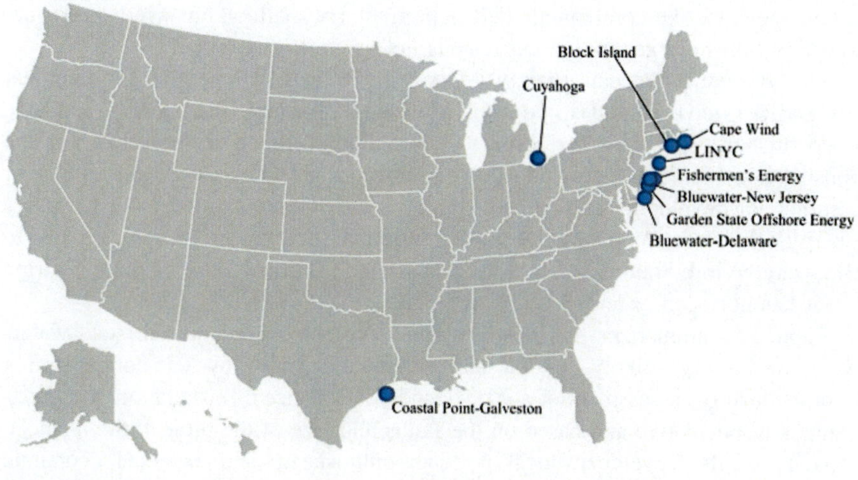

Fig. 1.4 Selected U.S. offshore wind projects *Source* [2]; industry press

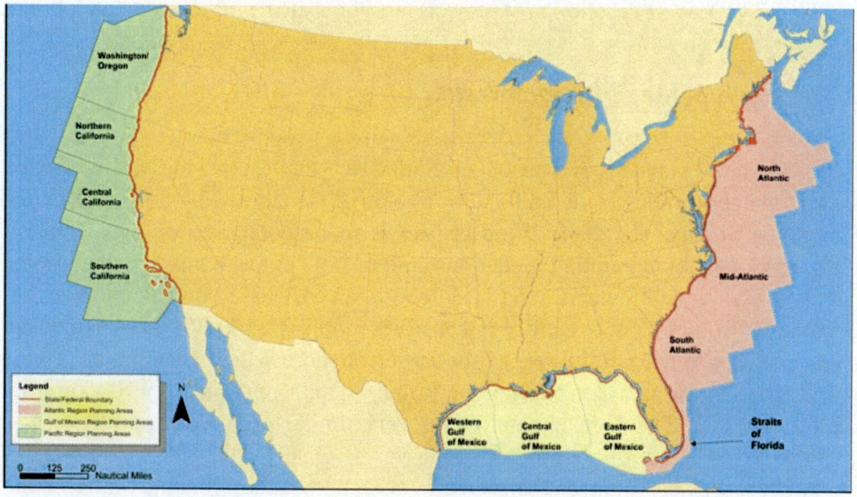

Fig. 1.5 BOEMRE planning areas *Source* BOEMRE

Demonstration and commercial projects are planned in both state and federal waters. Large-scale developments in state waters are most advanced in Texas, but significant development could also occur in the Great Lakes. One of the main obstacles for state offshore development is the aesthetic buffer that coastal residents may demand. For most states, state jurisdiction ends 3 nautical miles from the shore which may not allow for a sufficient buffer. Most development is expected to occur in federal waters, particularly the Northern and Mid-Atlantic regions (Fig. 1.5) where wind speeds are highest, state renewable portfolio standards exist, high population densities are found, and capacity limitations and transmission bottlenecks occur. The South Atlantic region has a geographic advantage since the continental shelf is particularly shallow, but wind speeds are lower and onshore generation capacity is not as constrained.

The Bureau of Ocean Energy Management, Regulation, and Enforcement (BOEMRE) have issued three limited leases offshore New Jersey and one lease offshore Delaware for resource evaluation. Limited leases allow for the construction of met towers and other monitoring activities but do not allow for commercial development. The New Jersey leases were awarded to Fisherman's Energy, Bluewater Wind, and Deepwater Wind. The Delaware lease was issued to Bluewater Wind. Southern Company was offered a limited lease offshore Georgia but it has not been executed [2].

Table 1.2 summarizes the most advanced projects in federal and state waters which we classify as likely, possible, and unlikely to be online by 2015 corresponding to our subjective assessment of a 90, 50, and 10% chance of installation. The judgments are speculative and based on the status and size of the project, the financial capacity of the developer, local political enthusiasm, and expected economic conditions. Negative movement in any of these factors will impair progress in development. We estimate that nameplate online capacity in 2015 will be on the order of 887–1,287 MW. A description of projects most likely to begin construction follows.

1.2.1 Cape Wind: Massachusetts

Cape Wind is planned in Nantucket Sound on the south side of Cape Cod approximately 6 miles from land and in a relatively shallow (3–15 m) area called Horseshoe Shoals. The Cape Wind project is to be composed of 130, 3.6 MW turbines placed on monopile foundations, and in 2011 Cape Wind signed a turbine supply contract with Siemens, and Middleboro, an MA-based manufacturing firm, is planning to build a new facility to construct monopiles. At 468 MW nameplate capacity Cape Wind will be one of the larger offshore wind developments in the world. The staging area is to be Quonset, Rhode Island, approximately 100 km from the offshore site. The development plan is very similar to European projects [3].

Cape Wind was proposed in 2001 and was originally under the jurisdiction of the Army Corps of Engineers before being assigned to the Minerals Management Service in 2005. The Minerals Management Service was reorganized and renamed the BOEMRE in 2010. Cape Wind was approved for commercial leasing by the Department of the Interior in April 2010, 16 months after the Environmental Impact Statement (EIS) was completed. Cape Wind was not approved through the current BOEMRE regulatory process, but through an interim policy specifically designed to allow it and one other project (now defunct) to continue through the leasing process while final regulations were created.

Cape Wind signed a Power Purchase Agreement (PPA) with National Grid in 2010 which agreed to purchase 50% of Cape Wind's power for 15 years at a price of 0.207 \$/kWh (increasing by 3.5% per year and including renewable energy credits, RECs) and signed a second contract for the remaining portion of the electricity with the expectation that this contract would be transferred to another utility. The Massachusetts Attorney General objected to the PPA and brokered the negotiation of a new contract at 0.187 $/kWh which was approved by the Department of Public Utilities. The contract results in approximately $4 billion in revenue in the first 15 years of operation without inflation adjustment. The Department of Public Utilities objected to National Grid's purchase of the remaining 50% of Cape Wind's power, and as of late 2011, that power remains unsold.

In 2011, Cape Wind was seeking financing and beginning contracting. Cape Wind failed to receive a federal loan guarantee but Siemens has offered to provide debt and equity financing for the project.

1.2.2 Coastal Point Energy-Galveston: Texas

Coastal Point Energy (formerly Wind Energy Systems Technology) has plans to develop five wind farms offshore Texas. The company is composed of former engineers active in the offshore oil and gas industry and its sister company (CPE Development) is developing in-house foundation manufacturing, installation, project management, and design and engineering capabilities. Coastal Point and CPE Development intend to conduct the majority of development work for their own projects and to bid as a contractor on other projects [4].

All of Coastal Point's projects are in state waters offshore Texas and the most advanced of these projects is off Galveston. In 2005, Coastal Point completed the first commercial offshore wind lease in the nation with the Texas General Land Office for 12,000 acres and leased an additional 70,000 acres in 2007 [5]. The lease requires an initial annual payment during preconstruction activities of 0.9–1.25 $/acre and a royalty payment of 3.5–6.5% during operation. The term of the lease is 30 years.

The Galveston wind farm is proposed for an area 7 miles offshore in 50 feet of water. The wind farm will consist of 60, 2.5 MW turbines placed on jacket-like foundations and is projected to cost $360–$450 million. Installation techniques are expected to be similar to those used in the oil and gas industry. Coastal Point installed a wind monitoring tower in 2007.

Coastal Point submitted a Rivers and Harbors Act permit application to the U.S. Army Corps of Engineers in late 2008 for wind farm construction and have received a permit for the construction of the first turbine. Coastal Point plans on installing a 2.75 MW turbine on an oil and gas platform by the end of 2011, but to date, no activity has been initiated. Coastal Point has acquired a port facility near Matagorda, TX and a fabrication facility near Lafayette, LA and is developing financing and awaiting regulatory approval.

1.2.3 Bluewater Wind: Delaware

In November 2006, Delmarva Power issued a request for proposals for the construction of a new power plant in Delaware in response to actions by the state legislature. Bluewater Wind submitted a proposal for a 450 MW wind park located 11.5 miles from shore. In May 2007, Bluewater was selected to negotiate a PPA, and in 2008, signed an agreement with Delmarva Power for the purchase of at least 200 MW of offshore wind power with the option to build up to 600 MW. The final PPA sets a price of approximately 0.12 $/kWh including energy and RECs and gives Delmarva Power 3.5 RECs for every credit purchased.

Bluewater's preliminary plans involve the use of monopile foundations and installation techniques using elevating vessels. Bluewater has expressed interest in building three turbine installation vessels at the Aker Shipyard in Philadelphia. They have also been in discussion with the turbine installation firm A2SEA on options for operating a U.S. flagged vessel for development [6].

In 2009, Bluewater received a limited lease from BOEMRE to install meteorological instruments. In late April 2010, BOEMRE published a request for interest in the Bluewater site, the first step in determining whether there is a competitive or conflicting interest in the site prior to commercial leasing. BOEMRE received responses from two commercial developers (Occidental Development and Bluewater) and in April 2011, Bluewater executed a non-competitive commercial lease with BOEMRE. In June 2011, Bluewater delayed the installation of a meteorological tower due to the upcoming termination of the federal renewable energy commercial loan guarantee program. According to the terms of its contract, Bluewater was required to pay Delmarva Power $4 million by June 23, 2011, but requested and received an extension to December 31, 2011 and is currently evaluating whether it plans to go forward with the development.

1.2.4 Deepwater Wind: Rhode Island

In 2008, Rhode Island solicited bids for the development of a commercial scale offshore wind farm. Deepwater Wind (formerly Winergy) was selected in early 2009. Two projects are planned: a small, 8 turbine, 29 MW wind farm off Block Island in state waters, and a farm composed of 107, 3.6 MW turbines (385 MW) approximately 20 miles offshore in 40–45 m water depths. Deepwater Wind has proposed a novel[3] plan for installing turbines, but thus far, has not secured a limited lease from BOEMRE.

Deepwater negotiated a 20-year PPA with National Grid for the purchase of electricity from the Block Island wind farm. The contract price was 0.244 $/kWh but was rejected by the Rhode Island Public Utility Commission for being too costly. The Rhode Island legislature passed a law requiring the Public Utility

[3] Deepwater Wind plans on using 4-pile jacket foundations transported to site by barge. The complete turbine is expected to be assembled onshore (although one blade may be installed after clearing bridges) and skidded onto a specially constructed, H-shaped jackup barge. The barge will be elevated so that it is level with the top of the foundation and the turbine will be skidded over the foundation and secured.

Commission to re-evaluate the project against more lenient standards and obligating Deepwater to disclose its development costs and the PPA contains provisions for price reductions if costs are lower than anticipated. The Public Utility Commission approved the PPA in 2010 and legal challenges were denied by the State Supreme Court in 2011.

1.2.5 Garden State Offshore Energy: New Jersey

Deepwater Wind and PSEG Global, a sister company of New Jersey's major regulated gas and electricity provider, plan to develop the Garden State Offshore Energy (GSOE) project, a 350 MW facility 17 miles off New Jersey. GSOE began in response to a solicitation from the New Jersey Board of Public Utilities. The proposed technology is similar to that used in the Deepwater Rhode Island project. Deepwater Wind received a limited lease from BOEMRE and plans to install a mobile wind monitoring buoy on the site.

1.3 Factors Impacting U.S. Development

Offshore wind development in the U.S. is constrained by three main factors: financing and PPAs, regulation, and the supply chain. A number of companies are interested in offshore wind development, however, the high costs and risks make financing difficult. Financing will rely on the ability of developers to negotiate favorable PPAs. Recently negotiated PPAs have provided developers with high expected returns, but have been challenged by regulators and politicians due to cost pass-through to ratepayers. The success of PPA's will depend on local electricity markets, federal and state greenhouse gas policies and grid operators' expectations of future hydrocarbon costs.

Financing for offshore wind farms was expected to be eased by the U.S. Department of Energy's renewable energy loan guarantee program. This program reduced borrowing costs for developers and risks for investors. However, the funding for loan guarantees for commercial scale renewable energy projects ended in September 2011 and no offshore wind developers received guarantees. Therefore, it is likely that developers will require private sector financing.

One of the major impediments to the development of offshore wind will be the evolution of shale gas. Plentiful natural gas supplies reduce the costs of gas fired electricity and make it significantly more difficult for offshore wind developers to compete. Additionally, large reserves will change utilities' expectations of future electricity costs which may reduce their willingness to pay for offshore wind projects.

The rate of leasing by BOEMRE in federal waters and regulatory approval by the Army Corps of Engineers in state waters is uncertain. Cape Wind has been approved, but its lease was not processed through the new BOEMRE regulations. No federal commercial lease has been awarded via the current BOEMRE regulations and the time between the Notice of Interest and the offer of a commercial lease and the time between submission and approval of a development plan are unknown.

Offshore wind development cannot occur without a reliable and efficient supply chain. Turbine installation vessels may be particularly hard to procure and procurement is likely subject to the Jones Act. Assuming that the average time to install a 3.6 MW turbine is 1 week, approximately 2,000 boat days are required to install 1 GW of capacity. To install this capacity in 4 years, an average of at least two, and likely three vessels, would be required. In the near-term, many turbine components and subsea cables will need to be imported from Europe which will add to development costs.

1.4 Atlantic Wind Connection

In October 2010, Google and Good Energies, an energy investment firm, announced plans to build a 6 GW HVDC transmission line in the federal waters of the Atlantic Ocean called the Atlantic Wind Connection (AWC). Initially, the plans call for a 150 mile cable connecting northern New Jersey to Delaware at a cost of $1.8 billion. The total project would run 350 miles from New Jersey to Norfolk, VA and cost $5 billion. The stated purpose of the cable is to transport electricity from inexpensive markets in Virginia to more expensive markets in New York and New Jersey, and to bring offshore wind power to shore. If built, both the power capacity and the length would make it among the largest submarine cables in the world.

The cable would likely smooth out fluctuations in power output from wind farms built along the route, making offshore wind more reliable and valuable [7]. Additionally, the cable could transmit power from areas where electricity prices are low, to higher price areas, potentially making the economics of development more favorable.

However, the movement of offshore wind power from the area in which it is produced to another region may prove problematic for local stakeholders. The governors of several states (Delaware, Maryland, Rhode Island, New Jersey) have supported offshore wind power as a means of meeting state Renewable Portfolio Standards (RPS). If power generated off the coast of one state was sold to the grid operator in another state, the power would not count towards RPS goals and the public may be less likely to support offshore wind.

There is also some concern that the AWC could be used to move coal-fired electricity from Virginia to northern markets and that offshore wind developers will have to compete with this less expensive power. AWC proponents argue that even if the AWC is used to move coal-powered electricity to Northeastern markets, the overall impact of the AWC on carbon emissions will still be positive [8].

Finally, the decision for a wind developer to interconnect with the AWC is complex and may not be preferred in all cases.

References

1. EWEA (2009) Oceans of opportunity. European Wind Energy Association, Brussels
2. Musial W, Ram B (2010) Large scale offshore wind power in the United States: assessment of opportunities and barriers. National Renewable Energy Laboratory. Golden, CO. NREL/TP-500-40745
3. MMS (2008) Cape Wind final environmental impact statement. Minerals Management Service. Herndon, VA. 2008-040
4. Coastal Point Energy (2010) Introduction to Coastal Point Energy. http://www.millergroupofcompanies.com/presentations/CoastalPointEnergyPPT3.pdf. Accessed 29 Oct 2010
5. Bogo J (2010) Texas to Cape Wind; you're not first yet. Popular Mechanics, May 3
6. Prowse M (2006) A2Sea vessel support letter
7. Kempton W, Pimenta FM, Vernon DE, Collie BA (2010) Electric power from offshore wind via synoptic-scale interconnection. Proc Nat Acad Sci 107(16):7240–7245
8. Pfeifenberger JP, Newell SA (2010) Direct testimony of Johannes P. Pfeifenberger and Samuel A. Newell on behalf of AWC. Federal Energy Regulatory Commission, Dec 20

Chapter 2
Offshore Wind Energy System Components

2.1 Meteorological Systems

A meteorological mast (or met tower) is the first structure installed during the planning stages. The purpose of a met tower is to evaluate the meteorological environment and resource data within the project area. A mast consists of a foundation, platform with boat loading, meteorological and other instrumentation, navigational lights and marking, and related equipment (Fig. 2.1). A buoy may also be used.

A mast collects wind data at multiple heights by intersecting the wind with an anemometer to characterize the project area's meteorology. Sensors collect data on vertical profiles of wind speed and direction, air temperature and barometric pressure, ocean current velocity and direction profiles, and sea water temperature. The data from the meteorological mast serve to test power performance, perform due diligence evaluation, and facilitate estimates of operation maintenance management.

Permit authorizations for the installation of monitoring systems are obtained through the U.S. Army Corps of Engineers (Nationwide permits 5 and 6), the U.S. Coast Guard (private aids to navigation), and the BOEMRE (limited lease) or state leasing agencies.

2.2 Support System

The support system refers to the foundation, transition piece, and scour protection. The primary purpose of the foundation is to support the turbine. A transition piece is attached to the foundation to absorb tolerances on inclination and simplify tower attachment. Scour protection helps to ensure that ocean conditions do not degrade the mechanical integrity of the support system.

M. J. Kaiser and B. F. Snyder, *Offshore Wind Energy Cost Modeling*,
Green Energy and Technology, DOI: 10.1007/978-1-4471-2488-7_2,

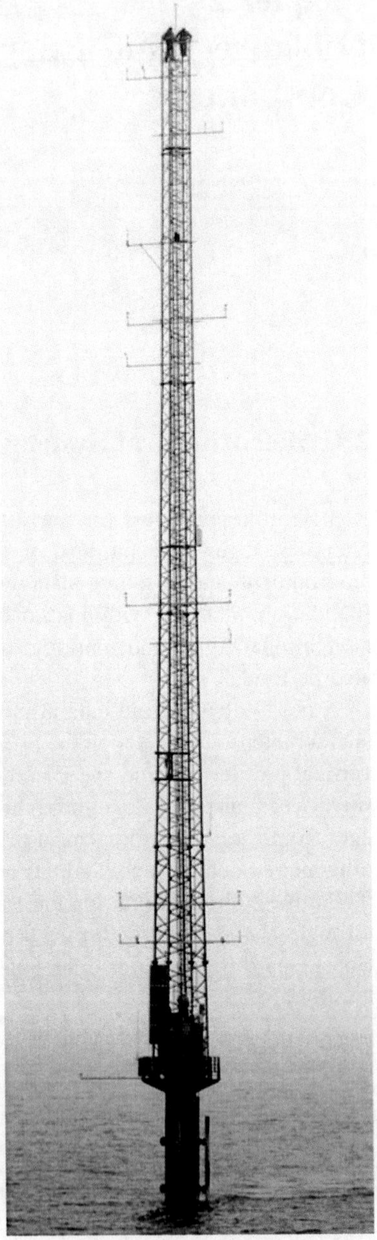

Fig. 2.1 Met tower at Horns Rev wind farm. *Source* DONG Energy

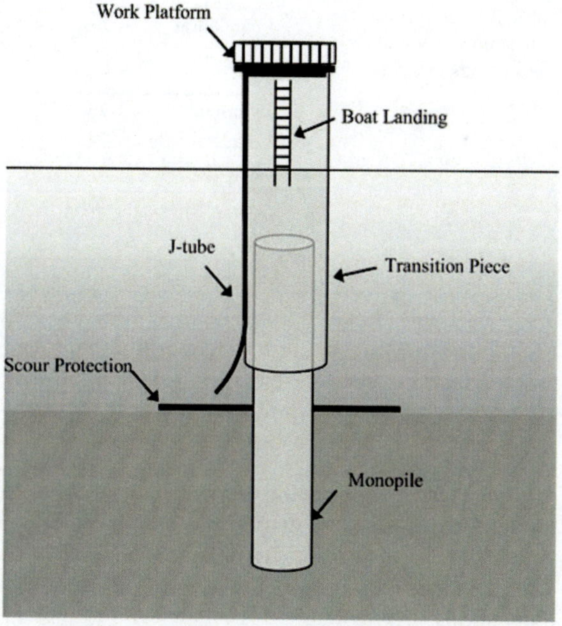

Fig. 2.2 Components of a monopile foundation

2.2.1 Foundation

Foundation technology is designed according to site conditions. Maximum wind speed, water depth, wave heights currents, and surf properties affect the foundation type and design. The size and weight of the turbine and tower are also key components. Within a wind farm, each foundation is customized to the water depth at its particular location.

Four basic types of foundations have been used in offshore wind farms: monopiles, jackets, tripods, and gravity foundation. Additionally, a single 2.3 MW demonstration turbine has been installed on a floating foundation. Foundations are prefabricated onshore in one piece, carried offshore by barge or other vessel, launched at sea, and set on bottom by a crane or derrick barge.

Monopiles

Monopiles are large diameter, thick walled, steel tubulars that are driven (hammered) or drilled (or both) into the seabed (Fig. 2.2). Outer diameters usually range from 4 to 6 m and typically 40–50% of the pile is inserted into the seabed. The thickness and the depth the piling is driven depend on the design load, soil conditions,[1] water depth, environmental conditions, and design codes. Pile driving is more efficient and less expensive than drilling. Monopiles are currently the most common foundation in shallow water (<20 m) development (Table 2.1) due to its

[1] In soft soil regions, deeper piles and thicker steel are required.

Table 2.1 Estimated distribution of foundation types of offshore wind farms

Foundation type	Installed by end of 2008 (%)	Planned for 2009–2011 (%)	Projected for 2011–2020 (%)
Monopile	75	80	50–60
Concrete base	24	15	5
Jacket/tripod	1	5	35–40

Source Bluewater Wind 2010

lower cost and simplicity, but because they are limited by depth and subsurface conditions, they are likely to decline in popularity in deeper water. However, in nascent markets such as the U.S., and for the near term future, monopiles are expected to be heavily employed.

Tripods
Tripods consist of a central steel shaft connected to three cylindrical steel tubes through which piles are driven into the seabed (Fig. 2.3). Tripods are heavier and more expensive to manufacture than monopiles, but are more useful in deep water. The Alpha Ventus project is the only operating wind farm that employs tripod foundations (Fig. 2.4).

Jackets
Jacket foundations are an open lattice steel truss template consisting of a welded frame of tubular members extending from the mudline to above the water surface (Fig. 2.5). Piling[2] is driven through each leg of the jacket and into the seabed or through skirt piles at the bottom of the foundation to secure the structure against lateral forces. Jackets are robust and heavy structures and require expensive equipment to transport and lift. To date, jacket foundations have not been used extensively due to the preference for shallow, near-shore environments. At around 50 m, jacket structures are required. Jackets have been used for two of the deepest developments, Beatrice (45 m) and Alpha Ventus (30 m), supporting large 5 MW turbines. Jackets are also commonly used to support offshore substations (Fig. 2.6). Jackets can be used in deep water (100s of meters), although economic considerations are likely to limit their deployment to water under 100 m.

Concrete Structures
Gravity foundations are concrete structures that use their weight to resist wind and wave loading (Fig. 2.7). Gravity foundations require unique fabrication facilities capable of accommodating their weight (either drydocks, reinforced quays, or dedicated barges). Gravity foundations have been used at several offshore wind

[2] Monopiles, jackets, and tripods are attached to the subsurface using piles. However, designs could be modified to use suction caissons in which a cylindrical steel caisson (resembling an overturned bucket) is allowed to sink into the seabed under its own weight [1]. Suction is applied to the inside of the caisson and water is pumped out. The resulting pressure differential causes the caisson to be driven into the seabed.

Fig. 2.3 Tripod foundations. *Source* Alpha Ventus

farms, including Middelgrunden, Nysted, Thornton Bank, and Lillgrund. Gravity foundations are less expensive to build than monopiles, but the installation costs are higher, largely due to the need for dredging and subsurface preparation and the use of specialized heavy-lift vessels (Figs. 2.8 and 2.9). The deepest gravity foundations in operation are in Thornton Bank (27 m). Gravity foundations are most likely to be used where piles cannot be driven and the region has dry-dock

Fig. 2.4 The Taklift 4 placing a tripod foundation at Alpha Ventus. *Source* Alpha Ventus

facilities for concrete construction [2]. Gravity foundations may also have an advantage in ice-prone regions [3].

In the North Sea, gravity foundations have been used in the offshore oil and gas industry, but in the U.S. there has been no use of concrete structures for offshore oil and gas operations and no plan to use them in offshore wind development. In Europe, gravity foundations will likely continue to fill an important niche for shallow to moderate water depth regions where drivability is a concern. However, they are unlikely to be used in U.S. waters.

Floating Structures
As water depth increases, the use of a steel platform will be limited by economic considerations. In the offshore oil and gas industry, the water depth limit for fixed platforms is about 450 m (1,500 ft), but in the offshore wind industry, the limit is likely to be less than 100 m because of economic conditions. Floating structures consist of a floating platform and an anchoring system. There are several alternative designs for floating turbine foundations all of which are variations on the spar and tension-leg concepts in the oil and gas industry (Fig. 2.10).

The Hywind concept is being developed by StatoilHydro. A pilot turbine was placed in waters off Norway in 2009 (Fig. 2.11). The foundation consists of an 8.3 m diameter, 100 m long submerged cylinder secured to the seabed by three mooring cables. Hywind was towed horizontally to a fjord and partially flooded and righted. Additional ballast was then added and the turbine installed on top. The assembled turbine was towed out to sea and the anchors were placed.

Fig. 2.5 A jacket foundation. *Source* Alpha Ventus

Blue H has developed a deep water concept based on the tension-leg platform. A prototype has been deployed off the coast of Italy and another is planned off the southern coast of Massachusetts. The Blue H concept consists of a two blade turbine placed on top of a buoyant, semi-submerged steel structure attached to a counterweight on the seabed. Plans are to assemble the turbine and foundation onshore and tow it to the offshore site.

Fig. 2.6 A jacket structure supports the substation at Alpha Ventus. *Source* Alpha Ventus

Fig. 2.7 Gravity foundations under construction for Thornton Bank. *Source* Luc van Braekel

Fig. 2.8 A gravity foundation being installed at Thornton Bank by the heavy-lift vessel *Rambiz*. *Source* Luc van Braekel

Fig. 2.9 The *Eide Barge* 5 lifting a gravity foundation from a barge at Nysted. *Source* DONG Energy

Fig. 2.10 The Hywind turbine and support structure. *Source* Statoil

2.2.2 Transition Piece

After the foundation is installed, a transition piece is placed on top of the foundation to create a level platform (Fig. 2.12). Transition pieces pass through most of the water column but do not rest on the seabed; boat fenders, access ladders, access deck, and handrails are attached on the outside. For monopile foundations, the gap between the pile and transition piece is normally filled with cement grout. For jackets and gravity foundations, transition pieces are installed in port and would not require a separate offshore lift, and do not contain boat landings, electrical conduits, or other accessory components as these are installed elsewhere on the foundation.

2.2.3 Scour Protection

When a structure is placed in a current and the seabed is erodible, scour may lead to structural instability. Scour refers to the removal of sediment from the area

Fig. 2.11 The Hywind turbine being towed offshore. *Source* Oyvind Hagen/Statoil

Fig. 2.12 Transition piece ready to be lowered on the monopile at Horns Rev II. *Source* DONG Energy

around the base of a support structure. Scour protection requirements depend on the current and wave regime at the site, substrate, and foundation type. Low tech and relatively inexpensive methods are usually adequate to address the problem. Commonly employed measures of scour protection include dumping rock of different grade and placing concrete mattresses around the foundation. For monopile foundations, a layer of small rocks may be installed prior to or following pile driving; later, after cabling is installed, large cover stones may be placed around the foundation [4]. Monopiles, gravity foundations, and tripods require significant scour protection, while piled jackets require little or no scour protection [5–7].

2.3 Wind Turbine

The wind turbine is composed of a tower, nacelle, hub, and blades. The blade/hub assembly is called the rotor. The tower is attached to the transition piece, and the nacelle is attached to the tower; the rotor is attached to the nacelle (Fig. 2.13). There are several different options for installation which will be discussed in Chap. 5.

Offshore turbines range from 2 to 5 MW and typical weights are shown in Table 2.2. Component size and weight varies with the electrical capacity of the turbine, the rotor dimensions, and the selection of blade, hub, and nacelle material and equipment. Turbines are an established commodity but offshore technology is in the early stages of evolution and will continue to develop. In 2011, Vestas released plans for a 7 MW offshore turbine and Siemens installed a prototype 6 MW gearless model. Sway plans on installing a prototype 10 MW turbine in late 2012.

Tower

Towers are tubular structures consisting of steel plate cut, rolled, and welded[3] together into large sections. The tower provides support to the turbine assembly and the balance of plant components, including a transformer located in the base,[4] a yaw motor located at the top, and communication and power cables. The tower also provides a ladder and/or an elevating mechanism to provide access to the nacelle. In installation, tower sections are bolted to each other during assembly, or are preassembled at port. Tower height is determined by the diameter of the rotor and the clearance above the water level. Typical tower heights are 60–80 m giving a total hub height of 70–90 m when added to the foundation height above the water line. Tower diameter and strength depend on the weight of the nacelle and expected wind loads.

---

[3] Manufacturers purchase steel as hot-rolled plates which are cold rolled and welded using standard machinery.

[4] The turbine transformer is either located up tower in the nacelle or at the base of the turbine (down tower). Turbine transformers take the energy generated by the turbine and convert it to approximately 34.5 kV for connection with the collection system.

Fig. 2.13 An assembled rotor being lifted onto a nacelle at Nysted. *Source* DONG Energy

Table 2.2 Weights of commonly used offshore turbines

Turbine	Capacity (MW)	Blade length (m)	Tower (t)	Rotor (t)	Nacelle (t)
Siemens 3.6-107	3.6	52	180–200	95	125
Vestas V90-3 MW	3	44	100–150	42	70
Repower 5 M	5	61.5	210–225	120	300

Source Company data

Nacelle

The nacelle houses the generator and gearbox and monitors communications, control, and environmental maintenance of the equipment (Fig. 2.14). The nacelle is principally composed of a main frame and cover. The main frame is the element to which the gearbox, generator, and brake are attached, and must transmit all the loads from the rotor and reaction loads from the generator and break to the tower [8]. Nacelles are large units and typically the heaviest and highest lift and play an important role in determining installation vessel suitability. The relative size of a nacelle is depicted in Figs. 2.15 and 2.16.

Hub

The hub is a cast steel structure which transmits horizontal wind loads from the blades to the nacelle and rotational energy to the gearbox via a low speed shaft.

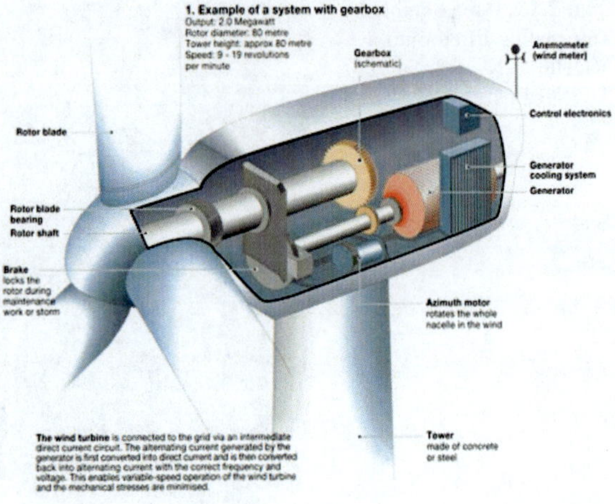

Fig. 2.14 Diagram of a nacelle. *Source* German Renewable Energies Agency

The hub is one of the most highly stressed components of the turbine [9] and may contain motors for controlling blade pitch.

Blades
Blades are airfoils made of composite or reinforced plastics. The blades are bolted to the hub either onshore or offshore. Due to the construction materials low weight and long length (50–60 m), blades are sensitive to high winds during lifting operations. The size and shape of assembled configurations complicate onshore and offshore transport.

2.4 Electricity Collection and Transmission

Cables connect the turbines and the wind farm to the electrical grid. Collection cables connect the output of strings (rows) of turbines depending on the configuration and layout of the wind farm. The output of multiple collection cables is combined at a common collection point or substation for transmission to shore (Fig. 2.17).

Inner-Array Cable
The inner-array cables connect the wind turbines within the array to each other and to an offshore substation if present. The turbine generator is low voltage (usually, less than 1 kV, often 500–600 V) which is not high enough for direct interconnection to other turbines. A turbine transformer steps up the voltage to 10–36 kV for cable connection. Inner-array cables are connected to the turbine transformer and exit the foundation near the mudline. Cables are buried 1–2 m below the mudline and connected to the transformer of the next turbine in the string. The power carried by cables increases as more turbines are connected and the cable size or voltage may increase to handle the increased load. Installation of

Fig. 2.15 Relative size of a nacelle. *Source* Siemens

Fig. 2.16 Relative size of a nacelle. *Source* Siemens

connection cable is performed in discrete steps from turbine to turbine. The amount of cabling required depends on the layout of the farm, the distance between turbines, and the number of turbines.

Export Cable

Export cables connect the wind farm to the onshore transmission system and is typically installed in one continuous operation. Export cables are buried to prevent exposure, and in some places, may require scour protection. At the beach, cables

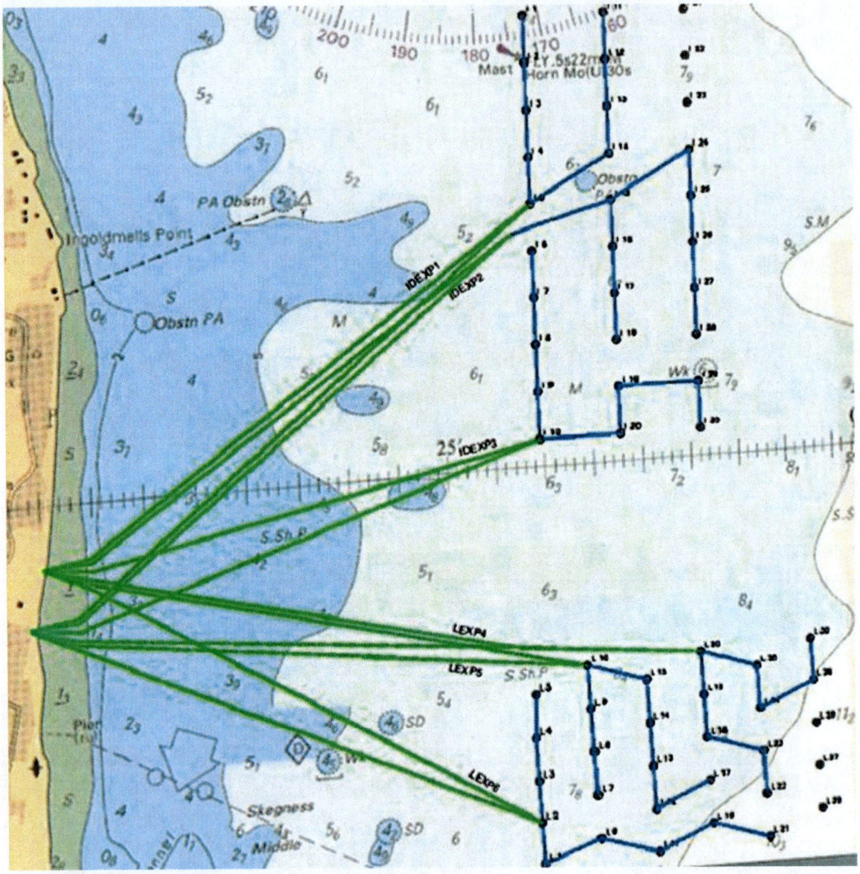

Fig. 2.17 Inner-array and export cable layout at Lynn and Inner Dowsing. *Source* Siemens

come onshore and may be spliced to a similar cable and/or connected to an onshore substation. Water depths along the cable route, soil type, coastline types, and many other factors determine the cable route, time, and cost. At the onshore substation or switchyard, energy from the offshore wind farm is delivered to the power grid. If the point of interconnection (POI) voltage is different from the submarine transmission, transformers are used to match the POI voltage; otherwise, a switchyard is used to directly interconnect the wind farm. At this point, power generated is metered and purchased via a PPA with a local utility or by entering the Independent System Operator's merchant market.

Export cables are composed of three insulated conductors protected by galvanized steel wire. Medium voltage cables are used when no offshore substation is installed and usually range between 24 and 36 kV. High voltage cables are typically 110–150 kV and are used with offshore substations. High voltage cables have the capacity to carry more power than a medium voltage cable but are heavier and

Fig. 2.18 Substation being lifted onto monopile at Gunfleet Sands. *Source* Offshore Wind Power Marine Services

wider in diameter. High voltage cable may weigh 50–100 kg/m while medium voltage cable may weigh 20–40 kg/m.

2.5 Offshore Substation

The purpose of an offshore substation is to increase the voltage of the electricity generated at the wind turbine to minimize transmission losses. The substation is sized with the appropriate power rating (MVA) for the project capacity, and steps up the line voltage from the collection system voltage to a higher voltage level, usually that of the POI.

All offshore wind farms require substations but not all substations are located offshore. The need for offshore substations depends upon the power generated and the distance to shore which determines the tradeoffs between capital expenditures and transmission losses [10]. The components of offshore substations include voltage transformers, switchgear, back up diesel generator and tank, accommodation facilities, j-tubes, and medium- and high-voltage cables. Substations are positioned within the wind farm at a location that minimizes export and inner-array cable distance. Substations are typically 500 tons or more and are placed on foundations similar to those used for turbines (Fig. 2.18). Onshore substations also include equipment to monitor power quality, such as voltage stability and harmonic disturbances, and SCADA systems allow the behavior of the entire system to be monitored and controlled.

2.6 Commissioning

Commissioning refers to the activities after all components are installed but before commercial operations begin. This includes electrical testing, turbine and cable inspection, and related quality control activities. The communication and control systems are tested to enable the turbine controllers to be accessed remotely from the control room.

References

1. Byrne B, Houlsby G, Martin C, Fish P (2002) Suction caisson foundations for offshore wind turbines. Wind Eng 26(3):145–155
2. Volund P (2005) Concrete is the future for offshore foundations. In: Proceedings of Copenhagen Offshore Wind, Copenhagen, 26–28 Oct 2005
3. Volund P, Jorgensen LB, Gravesen H, Sorensen SL, Pedersen B, Lorenz RS, Miller R, Ostergaard S, Riber HJ, Pedersen J, Gjerding JB (2003) Ice loads on offshore wind turbine foundations. Energi E2 #15822
4. Gerwick BC (2007) Construction of marine and offshore structures, 3rd edn. Taylor and Francis, Boca Raton
5. Den Boon JH, Sutherland J, Whitehouse R, Soulsby R, Stam CJM, Verhoeven K, Høgedal M, Hald T (2004) Scour behaviour and scour protection for monopile foundations of offshore wind turbines. European Wind Energy Conference, London, 22–25 Nov 2004
6. Seidel M (2007) Jacket substructures for the REpower 5 M wind turbine. European Offshore Wind Conference, Berlin, 4–6 Dec 2007
7. Larsen JHM, Soerensen HC, Christiansen E, Naef S, Vølund P (2005) Experiences from Middelgrunden 40 MW offshore wind farm. Copenhagen Offshore Wind Conference. 26–28 Oct 2005
8. Manwell JF, McGowan JG, Rogers AL (2002) Wind energy explained-theory design and applications. Wiley, New York
9. Hau E (2006) Wind turbines: fundamentals, technologies application economics. Springer, Berlin
10. Wright SD, Rogers AL, Manwell JF, Ellis A (2002) Transmission options for offshore wind farms in the United States. In: Proceedings of the American Wind Energy Association Annual Conference, 1–12 2002

Chapter 3
Leasing and Stages of Offshore Development

3.1 Project Location

Offshore areas in the U.S. are subject to various federal and state authorities (Fig. 3.1). Which agency has lead responsibility for regulatory oversight depends on the location and type of project and more than one state or federal agency may exercise jurisdiction.

3.1.1 Baseline

The baseline is the boundary line separating the land from the ocean. States use various points to demarcate private and state ownership, but for the federal government, the baseline is defined as the mean lower low water level (the lowest low-tide point) along the coast as shown on official U.S. nautical charts. It is drawn across the mouths of rivers and the entrances to bays, and along the outer points of complex coastlines.

3.1.2 State Waters

Generally, offshore state waters cover the area from the baseline out to three nautical miles[1] (nm), although it is to 9 nm for the offshore Gulf coasts of Texas and West Florida, as well as Puerto Rico, and to three imperial nautical miles offshore

[1] 1 nautical mile = 6,076 ft or 1,852 m. 1 imperial nautical mile = 6,080 ft.

M. J. Kaiser and B. F. Snyder, *Offshore Wind Energy Cost Modeling*,
Green Energy and Technology, DOI: 10.1007/978-1-4471-2488-7_3,

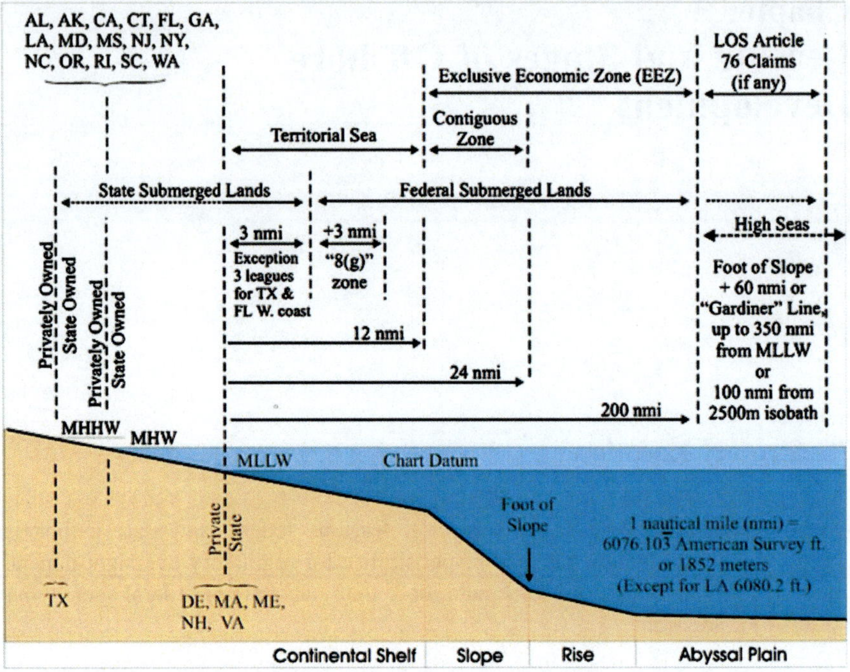

Fig. 3.1 Jurisdiction of U.S. offshore waters. *Source* NOAA

Louisiana. The U.S. half of the Great Lakes are state possessions. The area of state offshore jurisdiction was granted by the Submerged Lands Act of 1953 (43 U.S.C. §1301 et seq.). The federal government may regulate commerce, navigation, power generation (through the Federal Energy Regulatory Commission), national defense, and international affairs within this area, while states have the authority to manage, develop, and lease resources throughout the water column as well as on and under the associated seabed..

3.1.3 Outer Continental Shelf

The federal government administers the outer continental shelf (OCS), which comprises the submerged lands, subsoil, and seabed lying between the seaward extent of the states' jurisdiction and the seaward extent of federal jurisdiction. Typically, this is the area between 3 and 200 nm from the baseline.

3.2 Development Process

3.2.1 Lease Acquisition

The first step in developing an offshore wind farm is to acquire a wind lease from the landowner. For offshore wind, the landowner is either the state or the federal government, depending on the location of the site. On public lands, the government is responsible to ensure environmentally sound and safe development and a fair return to the public and all lease terms and provisions are publicly disclosed. On state and federal lands, offshore leases typically contain provisions for initial assessment periods and later commercial periods.

3.2.2 Assessment

After a company acquires a lease, the next step in development is to determine the wind potential at the site and conduct preliminary environmental studies. During assessment met towers are installed, engineering studies are performed, and the design process is initiated. If the property is considered commercially viable, financing is arranged, power purchase agreements are negotiated, and the capacity, model, and placement of turbines are determined. If the property is not suitable for development because of its wind resources or other impediments such as permitting problems, transmission interconnection issues, unfavorable power purchase price, or financing constraints, the wind lease terminates or the company does not exercise its option to develop. Commercially viable projects proceed to construction.

3.2.3 Design

Most design work occurs during the assessment phase. Offshore wind projects are engineered for the location in which they operate and because every location is unique a large degree of individualized design is required. Sites vary in terms of geological factors (seabed depths and morphology, seafloor composition, geology, seabed stability), meteorology and oceanography (wind, waves, currents, tidal range, seasonality), man-made factors (port infrastructure, shipping density, restricted areas, commercial fishing activities, oil and gas development, pipelines, submarine cables, buoys, dredging activity, shipwrecks, coastal construction, security threats), environmental factors (benthic, fish and shellfish, birds, marine mammals), landing site (utilities, geology, beach and near shore stability, weather exposure, environmental factors).

3.2.4 Construction

Construction activities include the procurement of goods, fabrication, assembly, and installation. Procurement is the largest cost category followed by fabrication, assembly, and installation. Construction normally includes the raw material cost (e.g., steel) and the cost to fabricate piling, transition pieces, and jacket structures, as well as the cost to assemble turbines. Fabrication may occur at the staging area or at another location, and depending on the degree of integration and pre-assembly required, one or more staging areas may be employed.

Procurement and Delivery
Procurement and delivery involves acquiring equipment and delivering it to the staging area. Equipment includes meteorological instruments, turbines (towers, nacelles, hub, blades), cable, transformers, supervisory control and data acquisition (SCADA) systems, fiber optics, and other goods that do not require fabrication. Turbines are the primary capital expenditure and are purchased according to a fixed cost and schedule of delivery. Turbines are delivered to the staging area and assembled according to the installation strategy. Turbines with the same generating capacity can have different weights, costs, and reliabilities, and developers will base selection on a combination of these factors. The capacity of the turbine will determine the number required and the model will set lift weights and vessel capabilities. Similarly, the decision to install high or medium voltage cable, and the voltage and weight of that cable will impact capital and installation costs.

Fabrication and Assembly
Foundations and transition pieces will be fabricated near or at the staging area from sheet steel, or may be delivered to site via road, rail, or barge. Turbine assembly activities may occur onshore or offshore and the degree of onshore assembly will impact installation costs. By reducing the number of offshore lifts, the time required by the installation vessel will decline but the lifting capacity requirements will increase. At the extreme, a complete turbine assembly will require one lift using a heavy-lift vessel.

Installation
Installation is defined as all activities to transport and install wind farm components from the onshore staging base to the onshore transition of the export cable. After a suitable number of turbines and/or foundations are available at the staging area, offshore installation may begin; however, if supply disruptions and delays occur, this will have significant negative consequences on installation expenditures. Installation occurs over three primary stages according to the unit being installed: (1) foundation, (2) turbines, (3) cable.

One or more marine vessels are required and the installation activities may be performed together or independent of one another. The basic work flow is as follows:

(1) Installation of the foundation, transition piece, and scour protection;
(2) Erection of the tower and turbine;

(3) Installation of the electric service platform (if applicable);
(4) Installation of the inner-array cables and scour protection;
(5) Installation of the export cables and onshore transition.

Installation activities include the cost to load the foundations, transition pieces, tower, turbines, substation, and cables from port; the cost to transport and install/erect individual elements; and all the support vessels directly associated with the operation (e.g., crew boats, tugs, and supply vessels). Offshore substation installation costs are included, but after cables reach shore, the costs to build (or upgrade) an onshore substation and onshore grid connection may or may not be included depending on the contracts with the grid operator.

3.2.5 Commissioning

After a wind farm is tested and commissioned, it is ready for commercial production and power injection into the grid.

3.3 Texas Offshore Wind Lease Terms and Conditions

3.3.1 General Conditions

Authority
The commissioner of the Texas General Land Office (GLO) grants leases of state-owned lands for wind development. Money earned by the GLO is constitutionally dedicated to the Texas Permanent School Fund.

Granting Clause
The granting clause provides the lessee the exclusive right to research and develop wind resources, to convert wind resources on the leased premises into electricity, and to collect and transmit that electricity to market.

Phases of the Lease
The terms and conditions of offshore leases are defined in terms of two phases: (I) research and analysis and (II) development and production [1]. Phase I is the research and analysis phase of the lease, where under an approved research plan, the lessee studies the wind resources at the site through meteorological towers, and performs other tests, such as soil and water samples, to determine the feasibility of commercial wind development. Phase I is of fixed duration and may be extended at the discretion of the GLO. Environmental studies on avian and bat interaction and migration patterns are required to be performed during this time, and the lessee is required to submit progress reports on a quarterly basis and a final report. At the end of Phase I, the lessee has an option to submit a production plan or to terminate the lease. In the case of termination, the lessee will need to remove all meteorological towers and instrumentation facilities.

Phase II of the lease is the construction and production phase, in which an approved construction plan is executed and wind electricity is generated. The duration of the construction period and the option to extend depends on the size of the proposed facility and other factors specific to the lease. The production period is typically for 30 years as long as there is commercial production of wind generated electricity. At the end of the production period, the lessee may enter into negotiations for a lease extension. At the end of Phase II, or in the case of default or termination, the lessee is required to return the greenfield status of the area.

Reservations
The full use of the leased premises are reserved to the State of Texas. Additional use of the lease may include recreational activities or exploration and production of minerals. The lessee's approved production plan contains the amount of acreage the lessee will retain for the project. The lessee also agrees to coordinate plans and cooperate on activities to minimize interference with other operations.

Retained Acreage
After production, the lessee will retain only that acreage that has been developed for production of wind electricity. Each wind turbine will retain an agreed amount of acreage, along with the wind easement, but all acreage outside the retained boundaries shall be released.

Removal Deposits
The lessee is required to deposit a bond, an irrevocable letter of credit, or a cash deposit in an amount sufficient to cover the surface and subsurface restoration costs and the estimated removal of the meteorological towers, wind turbines, and other related improvements prior to initiating construction. In Phase I, the removal deposit is to be based on the estimated quantity or type of research equipment the lessee intends to install. In Phase II, the removal deposit is based on the estimated project capacity. In both cases, if the amount of research equipment or generation capacity exceeds the estimated quantities at the time the bond is invoked, an increase in the removal deposit may be required.

Reporting Requirements
Reporting requirements are made at every major milestone during each period. In Phase I, a research plan is submitted for review and approval before initiating any physical activities on the lease, consisting of: administrative information, projected schedule and discussion of research methods and activities, proposal areas of operations and placement of meteorological towers, wind resource data/information gathered, and plans outlining required environmental impact studies.

During Phase I, quarterly progress reports are made, including copies of all studies, data, surveys, or test reports compiled or completed during the period. Proprietary data are to remain confidential for a period of no less than 24 months from the delivery of the reports. At the completion of Phase I, a final report is to be submitted describing the lessee's schedule and method for removing meteorological towers. After delivery of the Phase I final report, the lessee may exercise its option to terminate the lease or submit a production plan for review and approval, which will include administrative information, projected schedule and discussion of proposed production activities, analysis of market potential, research plan test data, and development strategy. The development strategy includes information related to the number, size, and location of each turbine installed, the type and manufacture of the turbines, the amount of retained acreage required by each turbine, and the improvement expected to be made.

Phase II consists of a construction period and a production period. Quarterly construction and installation progress reports are to be submitted during Phase II up to the production commencement[2] date. A final construction report is to be submitted prior to production. At the end of Phase II, a final report describing the scheduling and method for removing the wind turbines and other improvements is to be submitted. The lessee is required to provide copies of the contracts under which electricity is to be sold, and all subsequent agreements and amendments to those contracts [2].

Assignment
The lease may not be assigned, nor the leased premises subleased, without the prior written consent of the lessor. A change in ownership in the lessee's business entity in excess of a specific amount is subject to approval.

Ownership of Improvements
The lessee shall own all improvements on the lease, but the lessor reserves the right to retain the lower section or platform structure that supports the wind turbines for creating or maintaining an artificial reef. The lessor may elect not to have the lower section removed, and in such case, the lessee will be excused from its obligation to remove the lower section.

Insurance
The lessee must carry policies of insurance and/or bonds in amounts specified by federal or state requirements, or under limits determined by the GLO, prior to the commencement of work and through the expiration of the lease. Required coverage and typical limits are as follows: workers compensation—statutory; general liability and comprehensive auto—$1 million per occurrence; officers and director's liability, maritime, and environmental pollution—reasonable limits.

3.3.2 Galveston Island Lease Terms

Location
The Coastal Point Energy-Galveston offshore wind farm leases are located on blocks 187L and 188L in the Gulf of Mexico and comprise 11,355 acres approximately 7 miles offshore. The effective date of the lease was March 2005 (Table 3.1).

---

[2] Depending upon the development plan, the construction period may overlap the production period.

Table 3.1 Lease terms for Galveston Island wind farm

Operator	Galveston Office Wind, L.L.C.	
Effective date	March 2005	
Lease size	11,355 acres	
Location	5–10 miles offshore Galveston Island	
Research period	2 years + negotiated extension	
Construction period	3 years + 2 years extension	
Production period	30 years	
Project capacity	Minimum 150 MW	
No. turbines × unit capacity	50 × 3 MW	
Annual rent	$10,000/year during research and construction period	
Production royalty	Production period (year)	Percentage of gross revenue[a,b] (%)
	1–8	3.5
	9–16	4.5
	17–30	5.5
Minimum annual royalty	Production period (year)	Minimum royalty[c] ($ per MW installed)
	1–8	616,000
	9–16	836,000
	17–30	1,145,800
Royalty in-kind	Yes	
Force majeure and qualified interruption conditions	Applies to construction phase, production phase, and minimum annual royalty payments	
Removal deposit	Applies to research equipment, wind turbines, and/or related improvements in an amount sufficient to cover restoration, removal, and other costs. May take the form of a bond, letter of credit, or cash.	

Notes [a] If commercial production commences more than 40 months after the beginning of Phase 2, the production royalty is increased to 3.85, 4.95, and 6.05%, respectively

[b] Gross revenue is defined as the total amount of money or other consideration received; e.g., ancillary environmental benefits, from the sale of wind-generated electricity

[c] A reduction in the minimum annual royalty is possible if there is a qualified interruption of commercial production

Research and Construction

The research period is for a 2-year term with an extension that can be negotiated. The construction period is for 3 years with a possible 2-year extension.

Annual Rent

An annual rent of $10,000 applies between the effective date of the lease and the production commencement date

Production Royalty
A production royalty is based on a percentage of gross revenue (GR) which varies with the year of production t, beginning from the production commencement date ($t = 0$), as follows:

$$\mathrm{ROY} = \begin{cases} 0.035\,\mathrm{GR} & 0 \le \mathrm{t} \le 8\ \text{years} \\ 0.045\,\mathrm{GR} & 9 \le \mathrm{t} \le 16\ \text{years} \\ 0.055\,\mathrm{GR} & 17 \le \mathrm{t} \le 30\ \text{years} \end{cases}$$

Gross revenue is defined as the total amount of money or other considerations[3] received from the sale of wind-generated electricity before deductions. For electricity produced by wind turbines that are attained more than 40 months after the beginning of Phase II, the production royalty is increased to 3.85, 4.95, and 6.05%, respectively.

Minimum Annual Royalty
A minimum annual royalty (MAR) applies based on the year of production:

$$\mathrm{MAR} = \begin{cases} \$\,616,000 & 0 \le \mathrm{t} \le 8\ \text{years} \\ \$\,836,000 & 9 \le \mathrm{t} \le 16\ \text{years} \\ \$\,1,064,000 & 17 \le \mathrm{t} \le 30\ \text{years} \end{cases}$$

and guarantees a minimum income to the state over the life of the lease. The lessee shall pay the positive difference, if any, between the MAR and the aggregate quarterly production royalty. The MAR does not apply to partial commencement production.

Qualified Interruption and Force Majeure
A lessee may qualify for a reduction in its MAR if there is a qualified interruption of commercial production, arising from mechanical failures, equipment shortages, or other similar industry shortcomings that are not the result of the lessee's negligence. In such cases, the MAR is replaced by the year's actual production rate.

If the lessee is prevented from complying with any express or implied covenant, by reason of war, terrorism, rebellion, riot, strikes, Acts of God, rule, or regulation of government authority, that are beyond the control of the lessee's obligation to comply with such covenant, shall be suspended. Hurricanes, storms, or collision with a vessel are typical force majeure events.

3.4 Federal Legislation

Section 388 of the Energy Policy Act of 2005 (EPACT), Public Law 109-58 (H.R. 6), enacted August 8, 2005, authorizes the U.S. Department of the Interior to grant leases, easements, or rights-of-way (ROW) on the Outer Continental Shelf for the

---

[3] Other considerations include ancillary environmental benefits, such as credits, credit certification, or similar items, including renewable energy certificates, but excluding federal production tax credits, investment credits, or other tax credits.

development of renewable energy and to allow for alternate uses of existing facilities on the OCS. The Outer Continental Shelf Lands Act (43 U.S.C. §1331–1337) stipulates that energy developers operating on the OCS are required to have a federal lease for their project. Originally limited to oil and gas resources, §8 of the Outer Continental Shelf Lands Act was amended by EPACT to include energy from sources other than oil and gas, naming the Department of the Interior as the lead agency.

EPACT stipulates that BOEMRE authority does not supersede the existing authority of any other agency for renewable energy project permitting. The other primary agency involved in permitting is the Army Corps of Engineers which, through Section 10 of the Rivers and Harbors Act, is tasked with determining whether an installation poses a hazard to navigation. BOEMRE was not granted jurisdiction over areas within the boundaries of the National Park System, national wildlife refuges, national monuments, or the National Marine Sanctuary System.

The National Environmental Policy Act of 1969 (NRPA, 42 U.S.C. §4332) stipulates that federal agencies must prepare an environmental impact statement (EIS) on major federal actions with potential for significant changes to the quality of the environment. BOEMRE has determined that establishing the alternative energy and alternate use (AEAU) Program and rulemaking constitute a major federal action and has completed a programmatic EIS on its proposed AEAU Program.

3.5 Federal Offshore Wind Lease Terms and Conditions

3.5.1 Types of Access Rights

The BOEMRE issues access rights for commercial development and site assessment and technology testing of renewable energy facilities. There are two types of leases (commercial and limited) and two processes for awarding leases (competitive and noncompetitive). The BOEMRE also issues right of use and easements (RUE) and ROW grants and alternate use RUEs.

A commercial lease provides the lessee full rights to apply for and receive the authorizations needed to assess, test, produce, and sell renewable energy on a commercial scale over the term of the lease. Competitive leases for full development and power generation include a 6-month preliminary term, a 5-year site assessment term, and a 25-year operations term. Non-competitive leases are similar but do not include the initial 6-month preliminary term.

A limited lease conveys access and operational rights for activities that support the production of electricity but not the sale, distribution, or other commercial use exceeding a limit specified in the lease. Limited leases may be issued for site assessment purposes only or for site assessment and development and testing of new or experimental technology. Competitive limited leases are issued for a 6-month preliminary term and a 5-year operations term and may be renewed but

Table 3.2 BOEMRE auction methods

Type of auction	Bid variable	Bidding process
Sealed bidding	Cash bonus or operating fee rate	One sealed bid per company per lease or packaged unit
Ascending bidding	Cash bonus or operating fee rate	Continuous bidding per lease
Two-stage bidding	An operating fee rate in one, both, or neither stage and a cash bonus in one, both, or neither stage.	Ascending or sealed bidding until: (i) only two bidders remain, or (ii) more than one bidder offers to pay the maximum bid amount. Stage two sealed or ascending bidding commences at some predetermined time after the end of stage-one bidding.
Multiple-factor bidding	Factors may include, but are not limited to: technical merit, timeliness, financing and economics, environmental considerations, public benefits, compatibility with state and local needs, cash bonus, rental rate, and an operating fee rate.	One proposal per company per lease or packaged bidding unit.

Source [3]

cannot be converted to commercial leases. Non-competitive limited leases are similar but do not include the initial 6-month preliminary term.

RUE and ROW grants authorize the use of a designated portion of the OCS and to allow for construction and use of cable for the purposes of gathering, transmitting, distributing, or otherwise transporting electricity.

3.5.2 Auction Method

There are several different methods that the BOEMRE may apply to auction leases (Table 3.2). Sealed bidding is the simplest and involves a single round of bidding by each participant. The bid variable is the cash bonus[4] or a recurring charge known as the operating fee rate.[5] The bid variable depends on the auction method and is the focus of competition, and if it satisfies the BOEMRE threshold criteria (reservation price) is used to rank and award winners. Operating fees do not apply to limited leases, and so in a limited lease auction the cash bonus is the only bid variable.

[4] In oil and gas leasing, BOEMRE employs a cash bonus and a fixed royalty rate.

[5] The operating fee rate is based on BOEMRE estimates of what the lease should produce, and is determined from the installed capacity, expected capacity factor, and regional power price. Unless specified otherwise, the operating fee rate is 2% at the time of commercial operations.

Table 3.3 Frequency of NEPA/CZMA reviews based on instrument held

Instrument held	BOEMRE process	NEPA documentation and CZMA review
Competitive Commercial Lease	Conduct lease sale and issue decision on plans	1. Lease sale and SAP 2. COP
Noncompetitive Commercial Lease	Negotiate and issue lease	1. Lease issuance and SAP 2. COP
Competitive Limited Lease	Conduct lease sale and issue decision on plan	1. Lease sale 2. GAP
Noncompetitive Limited Lease	Negotiate and issue lease	1. Lease issuance and GAP
Competitive ROW, RUE Grant	Conduct ROW, RUE sale and issue decision on plan	1. ROW, RUE sale 2. GAP
Noncompetitive ROW, RUE Grant	Negotiate and issue ROW, RUE grant	1. ROW, RUE issuance, and GAP

Source [4]

More complex forms of auction such as ascending bidding, two-stage bidding, and multiple factor bidding may also be employed if the BOEMRE decides that the methodology will work best under particular circumstances [3, 4]. The BOEMRE also reserves the right to reject any and all high bids regardless of the amount offered or bidding system used.

3.5.3 Leasing Process

To ensure that use of the OCS provides a fair return to the U.S. government, BOEMRE regulations establish a competitive process for issuing leases and grants. The leasing process follows a well-defined path that includes several notices and environmental regulations (Table 3.3).

Proposed Sale Notice

The areas available for leasing in a proposed sale is published in the Federal Register inviting public comment on the terms and conditions of the lease sale, including: lease size, lease term, payment requirements, performance requirements, and lease stipulations. The notice also provides information regarding the auction process and method, including: bidding procedures and systems, minimum bid, deposit amount, and award method. After the BOEMRE considers the public comments, a final sale notice is issued.

Noncompetitive Leasing

BOEMRE may issue a lease on a noncompetitive basis if it is determined after public notice of a proposed lease that there is no competitive interest. If the BOEMRE determines that there is no competitive interest, the lessee must submit a site assessment plan for review and prior to award of a noncompetitive lease.

Table 3.4 Commercial lease cash flows

Payment type	Amount	Timing of payment
Deposit (competitive lease)	20% of bid amount, or as specified	Deposit is paid when the bid is submitted
Acquisition fee (noncompetitive lease)	$0.25 per acre	Acquisition fee is due when submitting a request for a noncompetitive lease.
Balance of bonus	As specified in bid award	Due within 10 days of receipt of BOEMRE bid acceptance notice.
Financial assurance	$100,000 minimum	Due within 10 days of receipt of BOEMRE bid acceptance notice. Additional financial insurance must be provided before the approval of each of the SAP and the COP.
Rent	$3 per acre for the main lease area (greater of $5 per acre per year or $450 per year for easement acreage)	First 6 months' rent due within 10 days of receipt of BOEMRE bid acceptance notice. Subsequent rent payments due at the beginning of each subsequent one-year period.
Cost recovery fees	Case specific, depending on cost of preparing environmental impact statements, etc.	Periodic billing based on actual costs incurred during document processing.
Operating fee	Unless the fee rate is a bid variable, generally an operating fee rate of 2% would apply during the operations term.	Begins when COP is approved. Payments due as per the schedule specified in the lease of Final Sale Notice.

Source [4]

3.5.4 Lease Terms

Competitive and noncompetitive leasing terms are broadly similar. Competitive leases deposit 20% of their bid amount at the time of submission, and if the winning bidder, will pay the remainder shortly after being notified it has been awarded the lease. Non winners are returned the deposit. In a noncompetitive lease, an acquisition fee (typically $0.25/ac) must be paid when the lease request is submitted.

Competitive and noncompetitive lease holders pay the same rents and fees (Table 3.4). During the site assessment term of a commercial lease, annual rents of $3/ac must be paid in advance. Acreage used by easements require an annual rent of $5/ac (minimum of $450 per year) for the entire lease term. After BOEMRE approves a plan authorizing construction, and once commercial generation begins, rent payments stop (except easements) and operating fees begin.

3.5.5 Development Process

Once a company acquires a lease, ROW grant or RUE grant, it must submit plans to the BOEMRE for approval prior to development, construction, operations, and decommissioning. Three types of plans are required depending on the type of lease or grant held:

- Site assessment plan (SAP)
- Construction and operations plan (COP)
- General activities plan (GAP)

The SAP describes the activities a lessee plans to perform to characterize their commercial lease or to test technology devices. A COP is required before a lessee may begin construction and/or operations on a commercial lease. The COP describes the construction, operations, and conceptual decommissioning activities. A GAP is required before a lessee or grantee may begin activities on a limited lease or ROW or RUE grant. A GAP describes the meteorological and oceanographic data collection; technology testing; and construction activities, operations, and conceptual decommissioning activities..

3.5.6 Project Plans

Assessment and Activities Plans
A SAP for commercial leases and GAP for limited leases is a description of specific site characterization studies and plans for observation and measurement facilities. The SAP and GAP are similar in that lease applicants must describe their planned resource assessment and site characterization activities, including: met towers, geophysical and geological surveys, hazard surveys, archaeological surveys, and biological surveys.

Construction and Operations Plan
A commercial leaseholder must submit and receive authorization of COP before project buildout may begin. The COP describes the facilities required for the full commercial project and the construction, operations, and decommissioning processes. The COP will include design drawings, fabrication, and installation details of the transmission infrastructure, and information about required easements. The SAP expires once BOEMRE approves the COP and the operations term begins. The BOEMRE considers each SAP, GAP, and COP to be a plan subject to the requirements of the coastal zone management authority (CZMA).

Decommissioning Plan
A description of decommissioning activities is submitted in the SAP, GAP, and COP for environmental review and impact. The description will include anticipated removal methods, easement activity, waste disposal plans, transportation

options, and schedule. Initial decommissioning plans are conceptual, but more detailed plans are required prior to the expiration of the lease.

3.5.7 Decommissioning

All facilities must be removed to a depth of 15 ft below the mudline when they are no longer used for operations but no later than 2 years after the termination of the lease, ROW grant, or RUE grant. Facilities include turbines, foundation structures, pipelines, cables, and other structures and obstructions. Lessees and grant holders must verify clearance within 60 days after a facility has been removed.

Decisions on what facilities may remain, approval for alternative removal depths, partial removal, and reefing considerations are to be made on a case-by-case basis during the technical and environmental decommissioning review process. Lessees that do not comply with the BOEMRE-approved decommissioning plan risk the forfeiture of the project bond or other financial assurance, as well as civil and criminal penalties under Section 24 of the OCS Lands Act.

Similar to the offshore oil and gas industry that holds all current and previous working interest and record title owners responsible for site cleanup, all co-lessees and co-grant holders of renewable energy leases are jointly and severally responsible for meeting decommissioning obligations on their leases or grants.

References

1. Patterson J (2010) Development of offshore wind in Texas. Presented at the Workshop for Offshore Wind Energy Development, Charleston, South Carolina, 24 Mar 2010
2. Rhame B (2007) Texas offshore wind energy. In: Proceedings of Coastal Zone 07, Portland, Oregon, 22–26 July 2007
3. Melnyk MMW, Andersen RM (2009) Offshore power. PennWell, Tulsa
4. Department of the Interior, Minerals Management Services (2009) Renewable energy and alternate uses of existing facilities on the outer continental shelf. Fed Regist 74(81): 19638–19871 To be codified at 30 CFR Parts 250, 285, and 290 (MMS Final Rules)

Chapter 4
Offshore Project Characteristics and Cost Factors

4.1 System Design

4.1.1 Design Requirements are Site-Specific and Multi-Dimensional

Offshore wind projects are designed based on site-specific conditions, and as in all large-scale engineering development, a number of design factors are involved and numerous tradeoffs exist [1]:

- Water depth impacts foundation type and size.
- Wind regime and load capacities impact the configuration of the wind farm, selection of turbine capacity, and weather downtime during installation.
- Distance to shore impacts the need for offshore substations and the length of export cables.
- Site geology impacts foundation requirements and cable installation and dredging requirements.
- Wave conditions impact the foundation type and design and weather windows during installation.
- Currents drive sediment transport, determine scour requirements, and affect sea bottom characteristics.
- The type of onshore transition (sandy beach, marshland, hard rock environment, sensitive marine habitat) determines cable route, installation complexity, and vessel selection.

M. J. Kaiser and B. F. Snyder, *Offshore Wind Energy Cost Modeling*,
Green Energy and Technology, DOI: 10.1007/978-1-4471-2488-7_4,

Fig. 4.1 Layout of Lillgrund, Sweden. *Source* Vattenfall

4.1.2 Configuration is Dictated by Prevailing Wind Directions and Aesthetic Appeal

The configuration of a wind farm depends on the meteorological and oceanographic conditions; the size, number, and type of wind turbines; visual impacts; regulatory requirements; and various other site-specific conditions (Figs. 4.1, 4.2, 4.3). The prevailing wind resource is a primary determinant of the layout. Cluster design is often used to maximize turbine density in a given area and to minimize aesthetic impacts [2]. The total footprint of the wind farm is minimized to reduce inner-array cable requirements and the area of impact. Typically, turbines are spaced 10 rotor diameters apart in the direction of the prevailing wind and 5 rotor diameters apart perpendicular to this direction to avoid wake interference.

4.1.3 System Capacities Reflect Commercial Status

Offshore wind farms are described by several physical and system attributes including nameplate capacity, number and capacity of turbines, water depth, distance to shore, and foundation type. We classify the commercial status of projects according to nameplate capacity:

- Demonstration: <20 MW
- Pre-Commercial: 20–100 MW
- Small Commercial: 100–250 MW
- Full Commercial: 250–750 MW
- Large Commercial: >750 MW

Fig. 4.2 Layout of Middelgrunden, Denmark. *Source* Kim Hansen

Fig. 4.3 Layout of Alpha Ventus, Germany. *Source* Alpha Ventus

The first offshore wind turbines were installed in Sweden and Denmark as demonstration projects, and from 1992 to 1998, several prototypes[1] were installed

[1] Lely (1994, Netherlands), Tuno Knob (1995, Denmark), and Irene Vorrink (1996, Netherlands) were relatively small scale, located in shallow or sheltered waters, and were heavily engineered.

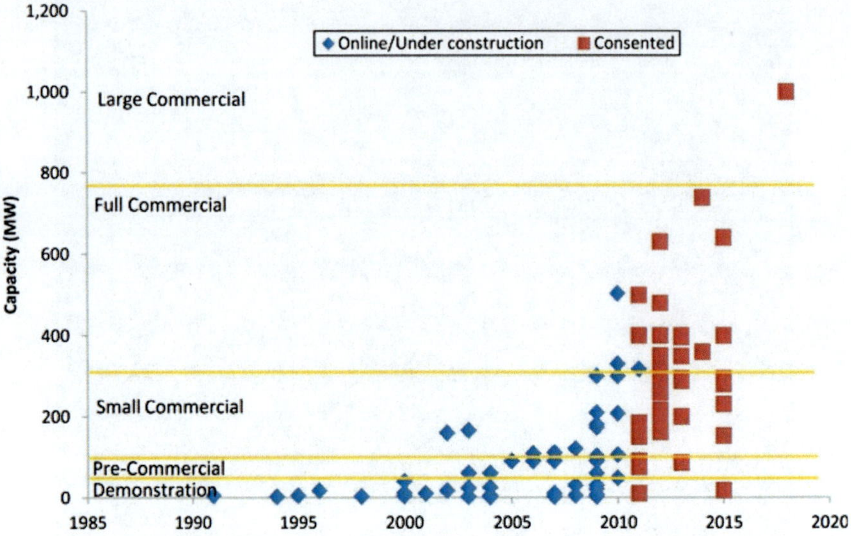

Fig. 4.4 Offshore wind farm capacity installation by year

throughout Europe. Demonstration projects usually involve less than 5 turbines and pre-commercial projects less than 20. The first pre-commercial projects were installed from 1998 to 2000. Commercial projects involve at least several dozen and perhaps as many as 150 or more turbines.

Figure 4.4 depicts the evolution of existing and consented European wind farms by nameplate capacity and installation date. The distribution of wind farms by class is shown in Fig. 4.5. In the U.S., there appears to be limited interest in developing demonstration or pre-commercial farms.

4.1.4 Turbine Selection has Broad Impacts on System Design

Offshore turbines currently range from 2 to 5 MW (Table 4.1). In early developments, 2 MW machines were common, but several recent projects have employed 5 MW turbines and 6 MW models are in testing. Currently, 3 and 3.6 MW turbines are the most common; the Siemens 3.6 MW turbine is the most popular offshore model followed by the Vestas V90 3 MW machine.

Turbine capacity determines the number of turbines and foundations required for a given system capacity and the power capacities of the inner-array cable system. Rotor diameter determines the turbine spacing and the length of the inner-array cable system. Turbine weight determines the type and size of the foundations and sets limits on the vessels capable of installing turbines.

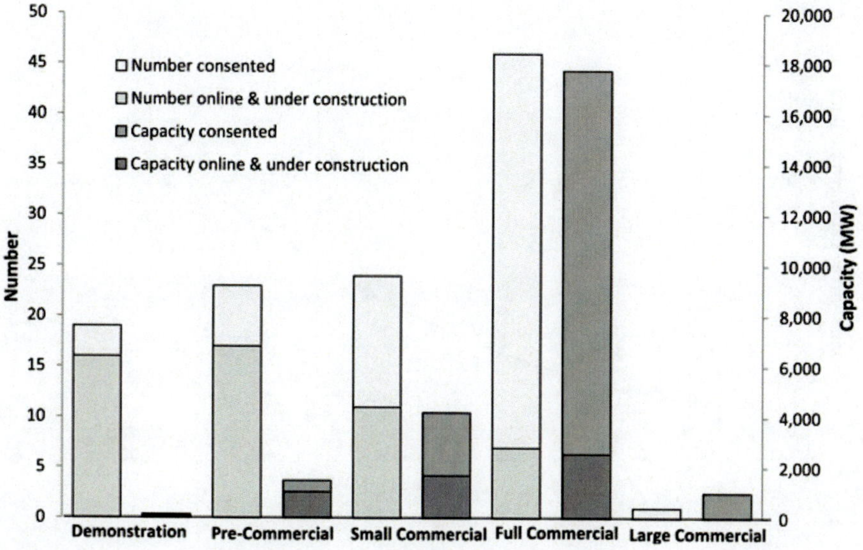

Fig. 4.5 Number and total capacity of wind farms by class

Table 4.1 Turbines used in offshore wind farms

Manufacturer	Model	Year available	Capacity (MW)	Rotor diameter (m)	Hub height (m)	Number installed[a]
Bard	5	2008	5	122	90	80
Multibrid	M5000	2005	5	116	90	6
Repower	5M	2005	5	126	90	44
Siemens	3.6	2005	3.6	107	80, 84	482
Siemens	2.3	2003	2.3	82	60–80	311
Vestas	V80	2000	2	80	67, 80	208
Vestas	V90	2004	3	90	80, 105	418

Note [a] Includes wind farms under construction as of October 2010
Source [12]

4.1.5 Common Physical and Engineering Requirements

Every wind farm is unique in configuration, design, and installation requirements, but the physical principles involved with extracting energy from the wind and the engineering requirements of design and installation create system correlations. For example, grid-like layouts are optimal in terms of energy generation and minimizing cable requirements, and this common physical feature will be observed through correlations between nameplate capacity, farm area, and cable length (Fig. 4.6). Rotor diameter and the number installed are also good predictors of farm area. System correlations reflect the physical engineering and economic relationships that are common to farm development.

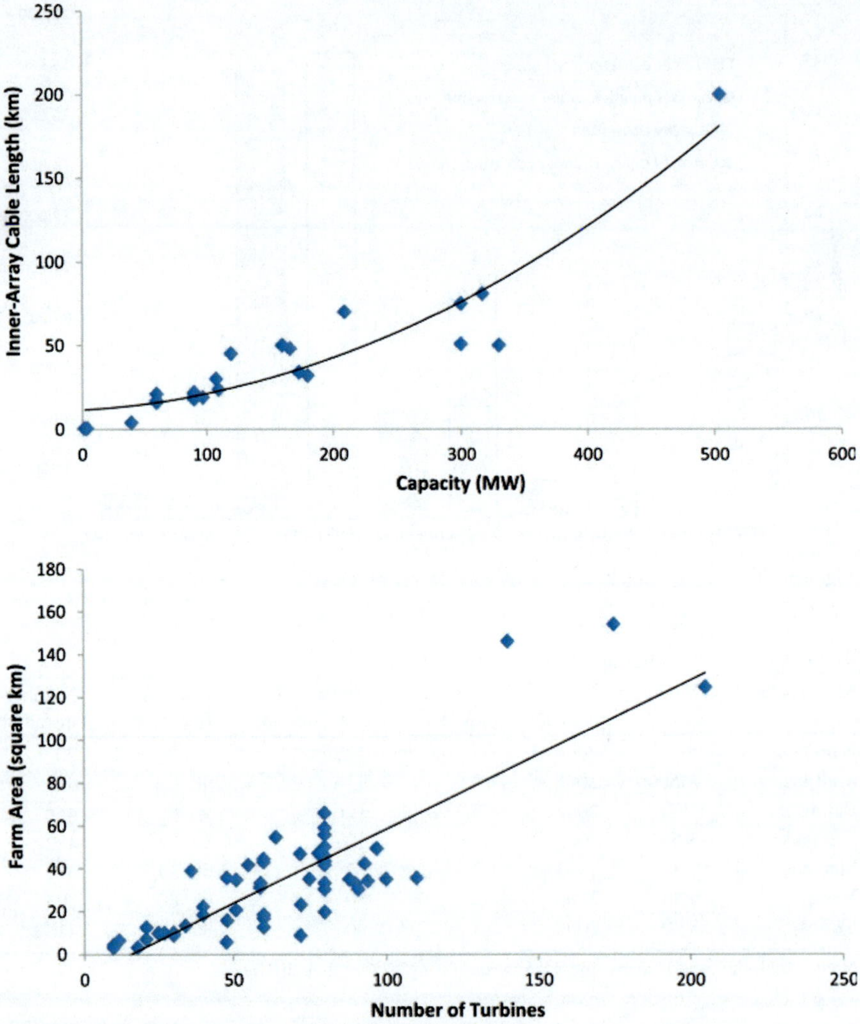

Fig. 4.6 System-level offshore wind farm correlations

4.1.6 Installation Cost Comparisons are Facilitated by System Homogeneity

System homogeneity allows conditions in one geographic region and installation space to be applied to another region and space. Homogeneity is realized not only through infrastructure correlations but is also manifested in installation requirements. Installation sequences and methods usually follow the same basic pattern with only minor modifications, and despite variations that may arise due to vessel selection and strategy, installation times are expected to remain reasonably

Table 4.2 Self-propelled installation vessel and jack-up barge tradeoffs

Jackup barge		SPIV	
Advantages	Disadvantages	Advantages	Disadvantages
Low dayrates	Large spread	Small spread	High dayrates
Available	Longer installation time	Limited supply	Typically shorter installation times
Simple to newbuild	Slow travel time	Fastest travel time	Expensive to newbuild
Useful in non-wind applications	Harder to maneuver into harbor	Increasingly popular in Europe	May be too expensive for non-wind applications
Lighter; may have more capacity	No dynamic positioning		
	Increasing capacity requires larger tugs		

consistent across developments. To date, only a small number of offshore turbine models and capacities have been available and these turbines have used geared generators causing the lift weights and foundation requirements to be similar across turbine models. In the future, as direct drive turbines increase in popularity and if more manufacturers enter the market, system components may become less homogenous. Firms have developed the most economic (shallowest) sites first using primarily monopile foundations. Monopiles require a specific installation sequence which creates system-level homogeneity.

4.2 Markets

4.2.1 Tradeoffs in Vessel Selection and Availability

Many vessel types and spreads can be utilized across each stage of the installation process, and generally speaking, owners seek the minimum cost at an acceptable risk from the fleet of vessels available and capable of performing the work requirements. In practice, a number of tradeoffs and constraints are involved in selection. Vessels that have less transport capacity have lower dayrates but require longer work times and involve a greater vessel spread. As the time required for installation increases, the risk of weather delays also increases. Table 4.2 shows some of the tradeoffs involved in the use of a high specification self-propelled installation vessel versus a lower specification jackup barge.

4.2.2 Installation Vessel Market Transparency is Poor

The supply and demand of marine vessels and levels of competition vary on a regional basis and this leads to differences in market rates, operator experience,

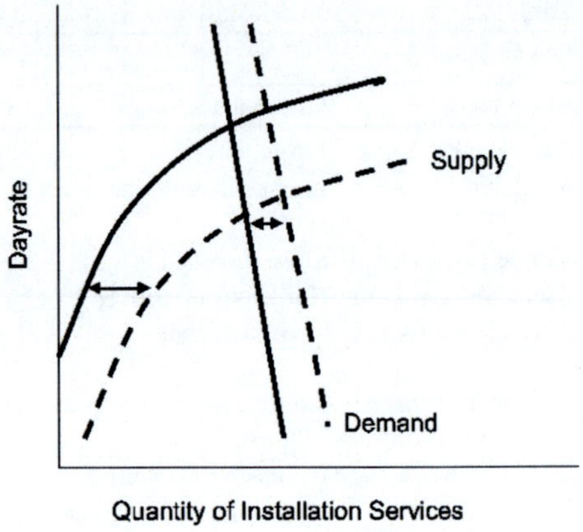

Fig. 4.7 Supply and demand factors influence installation vessel dayrates

risk, and the choice of contract types. Dayrates for offshore wind installation vessels are proprietary because of the small size of the market which is in contrast to the offshore oil and gas sector where highly competitive and transparent markets have evolved across most segments. The lack of market transparency adds to high levels of uncertainty in installation costs.

4.2.3 Varying Levels of Competition

Competition for marine vessels may come from within the wind industry or from other markets (civil construction, salvage, and oil/gas) and the degree of supply chain overlap depends upon the characteristics of the individual markets. Early European market development saw significant overlap, but as more specialized wind installation vessels are delivered the supply chain will diverge further.

In Europe, a large number of projects are at some stage of development, and with only a limited number of available vessels, dayrates have increased to allocate supply. Dayrates also send signals to investors in the newbuild market. In the U.S., only a limited number of vessels are available and demand is expected to increase gradually.

Figure 4.7 shows idealized supply and demand curves for turbine installation services in U.S. markets. Demand is set primarily by the costs of component supply, energy prices,[2] and government policy. Installation costs are a relatively small driver of overall demand, and as a result, the demand curve is relatively inelastic. At low prices the costs of vessel construction create barriers to market entry and supply is constrained. Once dayrates cover the costs of vessel construction, supply increases. Factors that shift the supply curve down include lower

[2] When oil prices are low, operators drill less wells and vessel utilization rates decline and vessels are available for other work. High price environments increase demand.

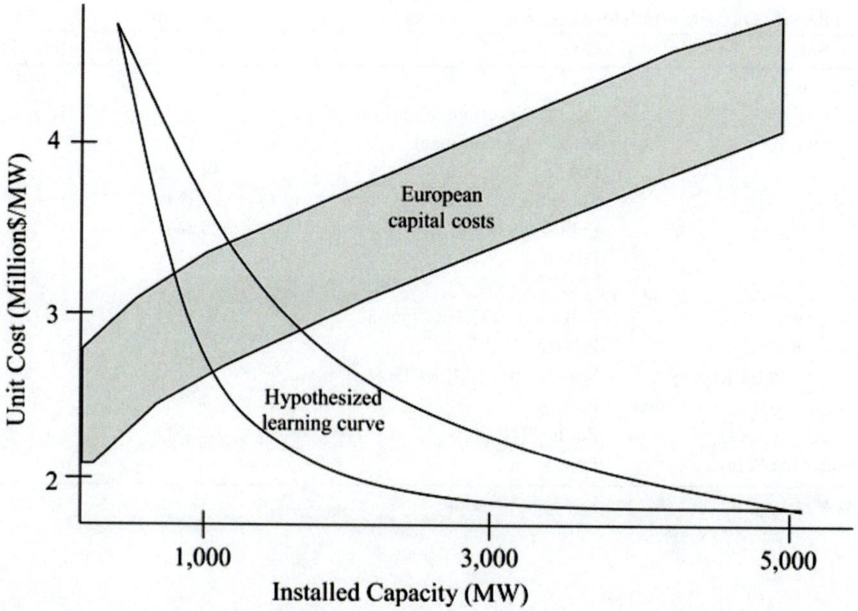

Fig. 4.8 Hypothetical learning curves for offshore wind and actual placement of the European market

costs for newbuilding, learning, or technological advancements in installation. Factors that move the demand curve to the right include government policies to address climate change, increasing costs of hydrocarbon-powered electricity, and reduced costs in other parts of the supply chain.

4.2.4 Learning Curve Uncertainty

In nascent markets such as the U.S., efficient operations are not likely to be achieved until regional competition and competencies mature, and in the short term learning effects are expected to be constrained. The shape of the learning curve is unknown but will impact costs in the near term (Fig. 4.8). As the capacity of offshore wind grows, costs are expected to decline due to learning effects on design, installation and project management, technological improvements, economies of scale, and market development [3, 4]. In Europe, capital costs for offshore wind have actually increased over the past decade which has been attributed to the high demand for turbine components and vessel services outpacing supply. As new turbine manufacturers and installation vessels enter the market, costs are expected to stabilize or decline. In the U.S., demand growth is expected to be more limited and costs may follow the hypothetical path, but costs are also expected to start higher as firms manage risk in a new market and develop manufacturing capacity and installation services under supply constraints.

Table 4.3 Offshore wind developers in Europe and the U.S.

Company	Ownership	Capitalization (billion $)
Europe		
DONG	Danish government and private equity	8 (E)
Vattenfall	Swedish government	20 (E)
E.ON.	Public	44 (M)
RWE Innogy	Subsidiary of RWE (public)	28 (M)
Centrica	Public	25 (M)
SSE	Public	16 (M)
U.S.		
Bluewater Wind	Subsidiary of NRG (public)	5 (M)
Deepwater Wind	Private	–
Coastal Point Energy	Subsidiary of Miller Group (private)	–
Energy Management Inc.	Private	–
PSEG	Public	17 (M)
Fisherman's Energy	Private	–

Note (E) equity and (M) market capitalization

4.2.5 U.S. Project Economics and Financing Strategies

Projects may be financed through recourse or non-recourse debt or may be self-financed. In recourse debt, the creditor may seek to recover money from the assets of the debtor; in non-recourse debt the creditor may seek possession of the wind farm, but would not be able to seek compensation from other assets. In Europe, most wind farms have used recourse debt or self-financing due to the financial strength of the project developers. In the U.S., project developers are largely small, private ventures formed specifically for offshore wind development. The U.S. companies will either need to seek equity partnerships with larger companies (as in the case of Bluewater Wind and the Garden State Project), or will require non-recourse debt, as their assets are unlikely to be sufficient for recourse loans (Table 4.3).

4.3 Contracts

4.3.1 Construction Contracts Define Cost Categories

Offshore wind farms have been developed under single Engineering Procurement Construction (EPC) contracts and multicontracts (Table 4.4). In an EPC contract, the developer solicits bids for the turnkey installation of a wind farm with a specified capacity and design. Under a multi-contract approach, development activities are tendered separately and it is the responsibility of the developer or project management firm to manage and negotiate the individual components. The

Table 4.4 Wind farm contract types

Windfarm	Contract type	Year online	Owners	Contractors
Horns Rev I	Multi	2002	Elsam, Eltra	Vestas, Nexans, A2SEA
North Hoyle	EPC	2003	Npower Renewables (RWE)	Vestas, Mayflower
Scroby Sands	EPC	2003	E.ON	Vestas
Nysted	Multi	2003	ENEGI E2, DONG, Sydkraft	ABB, Per Aarsleff, Bonus
Arklow	EPC	2003	GE, Airtricity	GE
Barrow	EPC	2005	DONG, Centrica	Vestas, KBR
Kentish Flats	EPC	2005	Elsam	Vestas
Burbo	Multi	2007	DONG	MT Hojgaard, Siemens
OWEZ	EPC	2007	Shell, Nuon	Ballast Nedam, Vestas
Princess Amalia	Multi	2007	Econcern, Eneco	Van Oord, Vestas
Lillgrund	Multi	2008	Vattenfall	Siemens, Pihl/Hochtief
Robin Rigg	Multi	2008	E.ON	Subocean, Areva, MT Hojgaard
Lynn/Inner Dowsing	Multi	2008	Centrica	MT Hojgaard, Siemens, Nexans
Horns Rev II	Multi	2009	DONG	A2SEA, Nexans, Siemens, Aarsleff
Rhyl Flats	Multi	2009	Npower Renewables (RWE)	MT Hojgaard, Ballast Nedham, Siemens
Alpha Ventus	Multi	2009	EWE, DOTI, E.ON, Vattenfall	Areva, REPower, Multibrid
Thornton Bank	Multi	2009	DEME, SRIW Ecotech Finance, EDF, SOCOFE, RWE, Nuhama	ABB, REPower, DEME
London Array	Multi	2012	E.ON, DONG, Masdar	MPI, A2SEA, Siemens, Nexans, Per Aarslef, Bilfinger Berger
Gunfleet Sands	Multi	2010	DONG	Siemends, MT Hojgaard, Pysmian, A2SEA, Ballast Nedam
Sheringham Shoal	Multi	2011	Statoil, Statkraft	Siemens, Nexans, Areva, Visser & Smit, MT Hojgaard, Master Marine
Walney	Multi	2011	DONG, Scottish and Southern	Ballast Nedam, Draka, Prysmian
Greater Gabbard	Multi	2012	Scottish and Southern, RWE	Flour, Siemens
Lincs	Multi	2012	DONG, Siemens, Centrica	MT Hojgaard, Siemens

manner in which contracts are written impacts how installation time and cost is reported.

Contracts may specify procurement, installation, and commissioning across one or more stages, either singly or in combination. Contracts that specify installation across individual stages (e.g., foundation, turbine, cable) provide the most direct evidence for installation cost since they are based upon well-defined activities. Unfortunately, for most contracts the cost associated with individual stages are not specified. Supply and installation activities are typically combined across one or more stages, and installation cost needs to be inferred from composite categories which creates additional uncertainty.[3]

4.3.2 Risk Allocation and Cost

In an EPC contract, the developer contracts with one company to provide services across all major work requirements (Fig. 4.9).

Total price is the bid factor and the level of competition at the time the contract is let plays an important role in determining cost. Risk is priced into services because a large number of uncertain activities need to be performed and it is the responsibility of the contractor to coordinate these activities and manage risk. EPC contracts have been reported to be 20% more expensive than similar construction via the multi-contract approach [5], but have the advantage of making project financing easier as lenders are assured of the project costs.

In the multi-contract approach, the developer will contract with several suppliers and installers for components and services (Fig. 4.10). Risk is not borne by a single party, but is spread between the developer and the contractor. As developers have gained experience they have become more capable of in-house project development [5, 6] allowing a shift from EPC to multicontracting [7]. Danish offshore wind farms have traditionally taken a multi-contract approach. Most new large offshore wind projects in the UK and continental Europe are developed using multicontracting.

4.3.3 U.S. Offshore Wind Will Likely Be Developed Using Multi-Contracting

Multicontracting will likely be the preferred approach in U.S. offshore development. EPC contracts are typically used when supply of installation services exceeds demand or when installation contractors have few alternative options [6].

[3] For example, an EPC contract might be written for the design, construction, and installation of foundations. In this case, the procurement, construction and installation costs are combined. Engineers may estimate the percentage attributable to each activity and report this value, but this may not be reflective of the actual cost breakdown. The only reliable method of determining project installation costs is through a detailed audit.

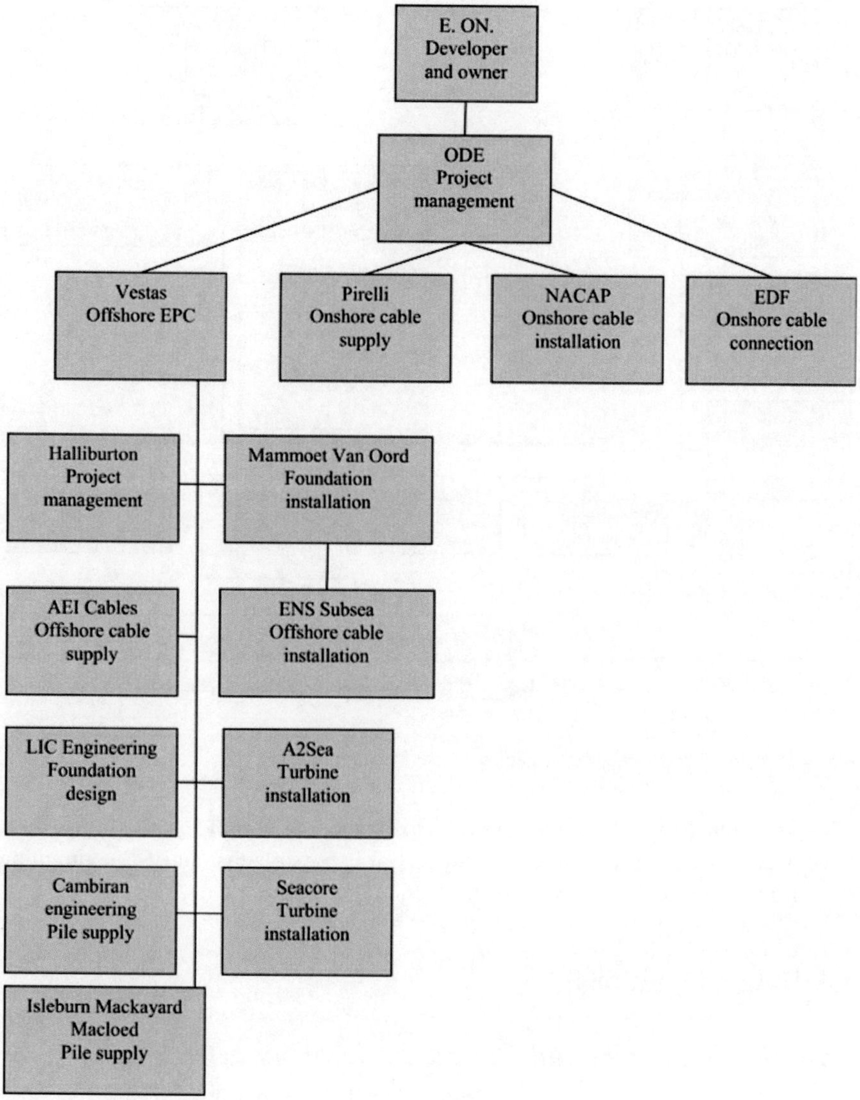

Fig. 4.9 Scroby Sands EPC contract structure. *Source* Modified from [5]

European contractors are unlikely to bid on U.S. EPC contracts if they have lower-risk options available in Europe. Indications from U.S. developers also suggest that a multi-contract approach will be favored. Cape Wind, Bluewater Wind, Deepwater Wind, and Coastal Point Energy have each conducted development activities that are not consistent with an EPC approach. These activities include signing supply contracts (Cape Wind), conducting detailed design work

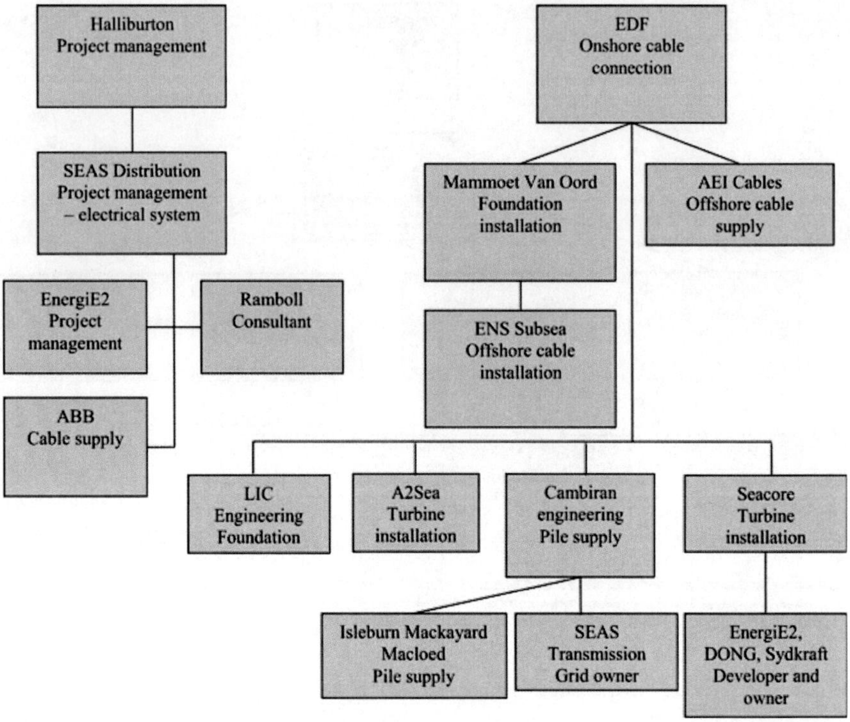

Fig. 4.10 Nysted multi-contract structure. *Source* Modified from [5]

(Coastal Point Energy), and attempting to finance or design installation vessels (Deepwater Wind and Bluewater). However, it is possible that an EPC contracting system may eventually develop in the U.S.

4.4 Data Limitations

4.4.1 Small Samples and Diverse Project Characteristics

In December 2011, there were 50 operating offshore wind farms in Europe with a total capacity of 4,555 MW and 4 projects (1,232 MW) under construction. These projects were developed over a period of 15 years and are of widely varying capacity. To compare projects with different characteristics, it is necessary to normalize across attributes meaning that it is necessary to group projects into categories for analysis. As additional attributes are used to categorize project data, the number of elements that populate each category declines, impacting the reliability of analysis and creating large variances in parameter estimation. Meaningful comparisons are therefore of limited utility in small samples with diverse characteristics.

4.4.2 Small Samples Limit Analytic Techniques

Small sample sets constrain empirical and statistical techniques and averages will be highly sensitive to outliers. Econometric techniques need to be applied with caution and normalized statistical measures must be interpreted carefully. Subjective assessment is necessary in the definition of reference (comparison) classes and assumptions on technology, learning, and inflationary factors. If new technological solutions are developed which result in a reduction in costs and execution time, these will not be reflected in approaches that rely on historical trends and current practices and technology. Over time, significant cost escalation has occurred in projects, but whether these trends continue into the future is uncertain, and how they impact U.S. projects is unknown.

4.4.3 No U.S. Projects are Under Construction

No U.S. projects have been constructed or are under construction through 2011. A number of offshore projects have been proposed for the East Coast, one project is in planning in the Gulf of Mexico, and another project is under consideration for the Great Lakes. Proposed projects and desktop engineering studies do not lead to a better understanding of cost drivers and factors impacting development since no direct experience has been gained and it is only through experience that learning and knowledge is achieved.

4.4.4 Apply Project Comparisons Cautiously

European markets are different from the U.S. market which limits the ability to extrapolate and compare projects and cost. The European offshore wind market has evolved with strong government support. Two of the largest offshore wind developers in Europe are partially owned by the state (DONG Energy and Vattenfall), and are likely to have different ways of valuing risk and the political benefits of offshore development. Furthermore, electricity and natural gas prices in Europe and the U.S. are different, with European natural gas prices tied to crude oil prices and thus significantly higher than the spot market prices in the U.S.

4.5 Cost

4.5.1 Vessel Dayrates are Market-Driven and Dynamic

The primary unknown variable in offshore construction cost is vessel dayrate. Competition levels and services are primary determinants of cost, and seasonal variations often exist. High competition stabilizes cost in normal markets, and supply and demand forces are a dominant factor impacting volatility. General inflationary pressures are also present. New firms can form relatively quickly and easily if demand and supply imbalances create the conditions for new business ventures to enter the market.

4.5.2 Impact of Catastrophic Failures

There are many risks associated with operating offshore. Catastrophic failure can occur due to extreme weather, design problems, or the impact of vessel collision, but the consequences are not expected to be as severe as in oil and gas operations[4] since there are no underground wells that need to be abandoned or flammable materials that pose a hazard. Wind turbines in the U.S. will likely be designed to withstand 100 year storms, however, this does not guarantee safety, especially given the duration of leases and the failure of many recently designed oil and gas structures over the 2004–2008 hurricane seasons [9]. The location of wind farms relative to shipping and tanker lanes increases the chance of collisions.

4.5.3 Port Facility and Location Impact Cost

Port facilities are critical for the offshore wind industry because they provide manufacturing facilities, bases for marine vessels, and staging areas to fabricate, assemble, and load-out the blades, nacelles, towers, transition pieces, and foundations necessary for development. There are a wide variety of port facilities in the U.S. and their ability to serve the offshore wind market depends on regional expertise, available space, and other factors.[5] The distance to the staging area determines the time to pick-up material and equipment at port and arrive at the work site. This will impact vessel spread requirements, work durations and installation costs. Port location relative to manufacturing sites and transportation networks also determines the transport costs of blades, nacelles and foundations. Foundations can be built at site or transported to the staging area by vessel, while most other components are expected to be transported over land.

---

[4] The risk and cost involved in decommissioning destroyed oil and gas infrastructure are higher than under normal conditions, often ranging between 5 and 20 times more than conventional abandonment [8].

[5] Desirable port characteristics generally include: deepwater, reinforced quaysides, large storage areas, easy access, suitable space, and facilities to move foundations.

4.5.4 Weather Risk is Common in Offshore Construction

Weather risk is an important factor in offshore work because it can delay operations and cause hazardous conditions. Installation is subject to wind speed, wave height, sea current, and tidal effects. Wind speed may hinder the installation of towers, nacelles, and rotor blades. High waves have the potential to disrupt any part of the installation. In multicontracts, weather risk is normally transferred to the individual contractors, while in EPC contracts, weather risk is accepted by the primary contractor.

Weather risk in the U.S. will be different from that experienced in Europe. Wind speeds along the U.S. coast are generally lower than those in Northern Europe, particularly along the U.S. Mid-Atlantic, South Atlantic and Gulf coasts. The Gulf of Mexico, and Atlantic coast are active hurricane corridors, and all sites are potentially at risk to extreme wind and wave stresses.

4.5.5 Public–Private Interface Impacts Cost Structure

State involvement can impact capital costs and significant differences exist between public involvement in Europe and the U.S. [10]. In addition to directly funding a portion of capital costs (i.e., the UK Capital Grants Scheme) or government ownership of developers (DONG and Vattenfall), government policies can impact installation costs through environmental and safety requirements, grid ownership, tax regimes, insurance requirements, and other policies. Even in cases where government policies do not directly impact capital costs, they may have indirect impacts. For example, the Production Tax Credit in the U.S. does not directly impact capital expenditures, but instead increases the value of the electricity produced. This can reduce risk for investors who may be willing to accept reduced rates, thereby reducing capital costs [11].

In the U.S., capital and installation costs may be influenced by the Jones Act and MARAD Title XI financing, both of which may impact vessel costs; the American Recovery and Reinvestment Act, which provides a 30% tax credit on renewable energy capital costs; the Production Tax Credit, which provides tax credits for every kWh of renewable energy produced; the National Environmental Policy Act, which mandates environmental reviews and may require mitigation techniques to be employed; state Renewable Portfolio Standards; local mandates for offshore wind development; administrative costs associated with regulatory compliance; and various labor laws concerning minimum and overtime wages and safety requirements.

4.6 Decommissioning

4.6.1 Decommissioning Timing is Defined by Regulatory Requirements

Energy companies that operate in federal offshore waters are obligated to remove all structures, clear the site, and verify clearance upon lease termination. Regulations for decommissioning renewable energy facilities and associated structures in federal waters are described in 30 CFR, Part 285, Subpart I, 285.900-913. All facilities, including pipelines, cable, and other structures and obstructions must be removed when they are no longer used for operations but no later than two years after the termination of the lease, right-of-way (ROW) grant, or right-of-use-and-easement (RUE) grant.

The scope of work for decommissioning in federal waters is determined by BOEMRE regulations. Endpoints for decommissioning are dictated by the lease instruments and specify a two-year timeline to conclude decommissioning which should allow operators to implement a cost-efficient means of decommissioning. However, decommissioning time requirements are not defined by capacity and therefore, a 150 MW wind farm has the same decommissioning timeline as a 750 MW wind farm.

4.6.2 Decommissioning Projects are Unique but Share Common Features

Each decommissioning project is unique in terms of the requirements of the operation, structure and site characteristics, equipment used, market conditions, contract terms, time of operation, and operator preferences. In oil and gas projects, decisions about when and how an offshore structure is decommissioned involve issues of environmental protection, safety, cost, and strategic opportunity. The timing and methods of removal are influenced by technical requirements as well as preference of the contractor and operator, scale economies, and scheduling [8]. In offshore wind farms, systems are more homogeneous and the operational requirements of decommissioning are expected to be more narrowly defined and similar across projects, leading to learning opportunities and cost reduction.

4.6.3 Decommissioning Operations are Low-Tech and Routine

Offshore decommissioning operations in the oil and gas industry are for the most part low-tech and routine involving standard equipment and procedures except in catastrophic situations. The physical requirements of decommissioning and the processes involved have not changed for decades and decommissioning operations are typically completed over time scales that range from a few days to several weeks. In offshore wind, decommissioning is also expected to be low-tech and routine, but will occur on a much larger scale; for example, one 300 MW decommissioning project with 100 structures will have activity levels similar to an entire season of Gulf of Mexico structure removals.

4.6.4 Learning Opportunities Will Occur

Offshore wind decommissioning projects in U.S. waters for installation that occur between 2015 and 2025 are expected to be similar: they will be composed of monopile or small jacket foundations in relatively shallow (under 50 m) water; projects will have a turbine tower and nacelle weight of 200–600 tons with a hub height of 65–90 m, and there will be 50 or more turbines for commercial projects and 20 or less turbines for demonstration projects. As a result, opportunities for scale economies and learning will occur.

4.7 Exposure and Liability

4.7.1 Joint and Several Liabilities

All co-lessees and co-grant holders of offshore energy leases in federal waters are jointly and severally responsible for decommissioning obligations. In the event that the operator cannot meet lease abandonment obligations, the responsibility of decommissioning would fall to co-lessees and working interest participants. According to BOEMRE regulations, "The BOEMRE looks first to the (designated) operator to perform these [decommissioning] obligations. Should the operator be unable to perform the lessee's obligations, BOEMRE will normally require any or all of the lessee(s) to perform." If there are no other co-lessees or working interest owners of the property, or if the current owners, either individually or collectively, are not able to perform, then any and all of the previous owners of the leases would be required to assume responsibility: "… If there is no lessee able to perform, BOEMRE will require prior lessee who held the lease during or after the time when the facilities were installed or the obstructions created to perform those functions."

4.7.2 Each Lease Represents a Different Level of Decommissioning Risk

From the operator's point of view, decommissioning activities represent a cost to be incurred in the future, while from the government's perspective, decommissioning represents an uncertain event and financial risk. Each offshore wind lease represents a different level of risk to the government which changes over time as properties age, as title holders and operating rights are transferred, and as the financial strength of companies change. The level of risk depends upon the number of current and previous record title holders and their financial capacity, the value of production relative to the cost of decommissioning, and the actual decommissioning expense relative to bonding requirements.

4.7.3 Bonding Protects the Public Interest

State and federal governments are exposed to financial liability if a company is unable to perform the requirements of the lease. The objective of a bonding program is to ensure that all entities performing activities under state or federal jurisdiction provide or demonstrate adequate financial resources to protect the government from incurring any financial loss. Securities help ensure that operators comply with all regulatory and lease requirements, including rents, royalties, environmental damage cleanup and restoration activities, decommissioning and site clearance, and other lease obligations.

4.7.4 Exposure Limits Vary With a Number of Factors

Government decommissioning exposure is determined by the size of the farm, the complexity of development, the financial strength of the company, and the level of bonding requirements. Ownership history will also play a role since all owners and co-owners are jointly and severally liable for decommissioning. In offshore oil and gas, it is common for properties to have multiple owners and to change ownership numerous times throughout the life cycle of the asset; it is not clear how financial transaction and ownership structures will evolve in offshore wind.

4.7.5 U.S. Government is the Party of Last Resort

The government is obliged to pay for decommissioning expense only if: the lessee cannot meet decommissioning obligations, there is no prior lease-holder capable of meeting obligations, the government cannot find a new lessee to take over the lease, and the cost of decommissioning exceeds the value of the bond. In the event that all these events occur simultaneously, government responsibility is determined as the cost of decommissioning less the bond value.

4.7.6 Bonding Cannot Provide Complete Protection from Noncompliance Risk

Bonding requirements are meant to provide financial assurance to the government that the owners of a lease will be able to return the property to its greenfield condition upon cessation of operation. Bonds are aimed at reducing—not eliminating—potential financial liability. Decommissioning bonds are meant to protect the government against incurring costs involved with removal and clearance activities and are set at a level that varies with risk tolerance. The costs of future decommissioning operations are always uncertain. If bond levels are set at ten times the expected cost of decommissioning, for example, the likelihood that the government would incur expense due to inadequate bonding would be a low-probability event, but the cost of doing business would be excessive which will reduce, and in many cases eliminate, market participants. There is a need for balance and rational expectations when setting bond levels [8].

4.7.7 Financial Failures in Offshore Wind May Be Less of a Threat

Offshore wind has a significant advantage over oil and gas development in terms of default risk to the government. In offshore wind, projects are developed with a Power Purchase Agreement which assures the operator a set price for produced electricity for the duration of the contract. As long as the utility (purchaser) remains a going concern, the wind farm does not suffer a catastrophic failure or significantly reduced output, an operator can be reasonably assured of a predictable income for a set period which reduces government risk. By contrast, in oil and gas production declines and commodity price fluctuation make revenue uncertain and decreasing toward the end of the project life.

4.7.8 All Bonding Procedures Have Limitations and Constraints

The adequacy of any bonding formula is based on the ability to estimate decommissioning cost accurately for a specific project and future period. All futures and cost estimates are uncertain because of project and market uncertainties and informational constraints. Bonding formula can only capture average characteristics under expected conditions; they cannot identify the specific requirements of decommissioning, what the service market will be at the time of the operation, what problems—if any—will arise in the activity, what options the operator has to reduce cost, etc. Bonding algorithms specify future decommissioning cost under current market conditions and technology and may be adjusted for inflation.

References

1. Manwell JF, Elkinton CN, Rogers AL, McGowan JG (2007) Review of design conditions applicable to offshore wind energy systems in the United States. Renew Sustain Energy Rev 11:210–234
2. Bishop ID, Miller DR (2007) Visual assessment of offshore wind turbines: the influence of distance, contrast, movement and social variables. Renew Energy 32:814–831

3. Bird L, Bolinger M, Gagliano T, Wiser R, Brown M, Parsons B (2005) Policies and market factors driving wind power development in the United States. Energy Policy 33:1397–1409
4. Lane N (2008) Issues affecting tidal, wave, and in-stream generation projects. Congressional Research Service, Washington RL 338833
5. Gerdes G, Tiedmann A, Zeelenberg S (2006) Case study: European offshore wind farms—a survey for the analysis of the experiences and lessons learnt. DENA
6. Hartley K (2006) The winds of change: EPC contracts for offshore wind farms. Monday Business Briefing. May 24, 2006
7. Penhale G (2008) Rhyl Flats offshore wind project delivery—a client's perspective. BWEA 30th Annual Conference, London, 21–23 Oct 2008
8. Kaiser MJ, Pulsipher AG (2008) A review and update of supplemental bonding requirements in the Gulf of Mexico. U.S. Department of Interior, Minerals Management Service, Herndon, VA. TAR No 600
9. Puskar F (2010) Assessment of damage and failure mechanisms for offshore structures and pipelines in Hurricanes Gustav and Ike. U.S. Department of the Interior, Bureau of Ocean Energy Management, Regulation and Enforcement, Herndon, VA. TAR No 642
10. Snyder B, Kaiser MJ (2009) Ecological and economic cost-benefit analysis of offshore wind energy. Renew Energy 34:1567–1578
11. Logan J, Kaplan SM (2008) Wind power in the United States: technology, economic, and policy issues. Congressional Research Service, Washington RL 34546
12. AWS Truewind (2009) Offshore wind technology overview. New York-Long Island Offshore Wind Collaborative, 9998

Chapter 5
Installation Strategies and Options

5.1 Foundation Installation

The methods for installing foundations are dictated by the foundation type. Within a foundation category, some variance exists in the methods employed, but the differences are relatively minor. We focus on monopiles and jackets as they are the most relevant to the U.S. market.

5.1.1 *Monopiles*

Monopiles may be transported to site by the installation vessel, they may be barged to the site, they may be transported by a feeder vessel, or they may be capped and wet towed (i.e. floated using their own buoyancy). The choice depends on the size and weight of the monopile, the variable deck load and crane capacity of the installation vessel, the distance to shore, and environmental conditions. Large installation vessels with heavy-lift cranes may be able to carry several monopiles from port and lift them into place. In contrast, smaller vessels with lower deck loads or low capacity cranes may require a wet tow or barge system.

After arrival on site the pile is upended so that it is sitting vertically on the seabed. This is accomplished by a crane and/or a specialized pile gripping device and is the step which usually defines the required crane capacity (Fig. 5.1). A hydraulic hammer is placed on top of the pile and the pile is driven into the seabed to a predetermined depth (Fig. 5.2). The time to drive the pile depends on the soil type, pile diameter and thickness, and the weight of the hammer. A rocky subsurface may prevent driving operations, in which case a drill will be inserted into the pile to drill through the substrate. At the conclusion of pile installation, the pile extends from several dozen meters below the mudline, to just above the water

M. J. Kaiser and B. F. Snyder, *Offshore Wind Energy Cost Modeling*,
Green Energy and Technology, DOI: 10.1007/978-1-4471-2488-7_5,

Fig. 5.1 Monopile being lifted off the deck. *Source* Chris Laurens/Vattenfall

line. The depth the pile is driven is determined by the soil type and design load, and typically about 30–50% of the total length is below the mudline.

After the monopile is secured in the seabed, a transition piece is lifted and grouted onto the pile (Fig. 5.3). The transition piece is typically installed immediately after piling by the same vessel that drove the pile, but if two vessels are employed in an "assembly-line" installation, a separate vessel may follow behind the monopile installation vessel and install the transition piece. The transition piece typically starts in the water column above the mudline and extends several meters above the top of the monopole. A side dumping barge or other utility vessel may be used to place rocks around the monopile to guard against erosion (scour protection).

5.1.2 Jackets and Tripods

Jackets and tripods are barged from the fabrication yard to the construction site and are lifted into place (Fig. 5.4). The piles used to secure jackets and tripods to

Fig. 5.2 Hammer placed on the top of a monopile before driving. *Source* Chris Laurens/Vattenfall

the seafloor are significantly smaller in diameter and length than monopile foundations because the mass of the jacket is better suited to hold the structure in place. Piles are either driven through sleeves at each corner of the foundation or the foundation may be placed over pre-driven piles. The sleeves are grouted to the pile or may be deformed to hold the jacket or tripod in place. The transition piece may be pre-attached to the foundation to save a lifting operation. For moderate water depths (30–50 m) jackets and tripods weigh between 500 and 800 tons. In the past, jackets have been installed by heavy-lift vessels, but some newbuild elevating vessels also have the lift capacity to place these foundations. Scour protection is less critical for jackets and tripods than for monopiles.

5.1.3 Factors Impacting Installation

Foundation type will impact the time required for installation. Jackets and tripods take longer to install than monopiles because they are heavier, more complex, and

Fig. 5.3 Transition piece being installed over monopile at Horns Rev 2. *Source* DONG A/S

more piles must be lifted and driven into place. Soil type can impact installation time because if hard rock exist below the mudline, piles will need to be drilled. If the surface is erodible, scour protection will be required which will increase the vessel spread and add time to installation. Soil type and maximum design loads also determine the required insertion depth to maintain a stable foundation.

If the installation vessel transports the foundations, the distance to port, vessel speed, and the number of foundations carried per trip determines the travel and loading time. Installation vessel travel time may be eliminated by transporting foundations on barges or towing them to site. The number of installations impacts both the total time and the time required per foundation.

Ideally, offshore work would occur in the seasons with the most favorable weather, however, this is frequently not possible and work often occurs in the winter where weather downtime is more common. Foundation installation is not as

Fig. 5.4 The *Thialf* placing a foundation at Alpha Ventus. *Source* Alpha Ventus

sensitive to wind conditions as turbine installation, but work during the winter will be associated with weather delays. In New England and the U.S. Mid-Atlantic, winter weather is not as severe as in the North Sea, but may still cause delays.

5.1.4 Proposed U.S. Foundation Installation Methods

U.S. foundation installation is expected to follow the European experience. Monopiles will be the preferred choice in shallow water and jackets or tripods will be the preferred choice in deeper water. Sites close to shore and in shallow water are less expensive and risky than deepwater sites and monopiles will predominate in the near term. Gravity foundations are unlikely to see any significant use in U.S. waters. It is possible that deepwater, floating foundations will one day become commercially and technically viable, but a considerable period of refinement is required before this will occur.

In the Cape Wind project, developers have proposed barging 3–4 monopiles per trip from Quonset, Rhode Island, 63 miles to Nantucket Sound. A jackup barge with crane will lift the monopiles from the transport barge and place them in position for pile driving operations. After the foundation is secured, a transition piece will be lifted and set atop the foundation and grouted [1].

5.2 Turbine Installation

Turbines are installed after foundations are in place. Installation may be done by the same vessel that installed the foundations or a different vessel may be used.

5.2.1 Transport

Most frequently, a single vessel both transports the turbine components and performs installation. Feeder vessels have also been used to transport turbine components, but this approach has not been as popular because of the risk associated with offshore transfers. When a feeder vessel is used, it often elevates to eliminate vessel movements. The decision to use a feeder vessel will depend largely on the transit speed and costs of the installation vessel, the deck load, the size of the turbine components, and the distance to shore.

5.2.2 Installation

There are a large number of options for turbine installation. When delivered, turbines typically consist of seven individual components, including three blades, at least two tower sections, the nacelle, and the hub. Some degree of onshore assembly is performed to reduce the number of offshore lifts and the degree of pre-assembly will impact vessel selection and installation time. Offshore lifts are risky and are susceptible to delay due to wind speeds, so preference is usually to maximize onshore assembly.

The methods used for offshore turbine installation are classified in terms of the number of lifts required as shown in Fig. 5.5

1. *Nacelle and hub joined onshore*. The tower is installed separately in two lifts, followed by the nacelle with the rotor hub preattached. All three blades are lifted separately. This method involves little onshore assembly and was used at Sprogo and Lynn and Inner Dowsing. At Lynn and Inner Dowsing this method was chosen because there was a long distance between the port and the offshore site and the method allowed for an efficient use of deck space permitting a large number of turbine components to be carried in a single trip.
2. *Tower assembled onshore*. The tower is assembled onshore and installed in a single lift. The nacelle and hub are lifted together and all three blades are installed separately. This method was used at Rhyl Flats and Burbo Bank (Fig. 5.6). As in Method 1, it has the disadvantage of requiring separate lifts for each blade which are susceptible to delays due to high winds. However, since the blades are not preassembled, they can be transported more easily, potentially allowing for a larger number of turbine components to be carried in the same deck space. Other than the deck area savings associated with not preassembling rotors, there is little reason to lift blades separately as the assembled rotor is unlikely to weigh more than the nacelle and would therefore not be the weight limiting lift.
3. *Rotor assembled onshore*. The tower is transported offshore in two pieces and lifted in two lifts. The nacelle is lifted separately. The rotor and all three blades are assembled onshore, transported to the offshore site, and lifted. This method

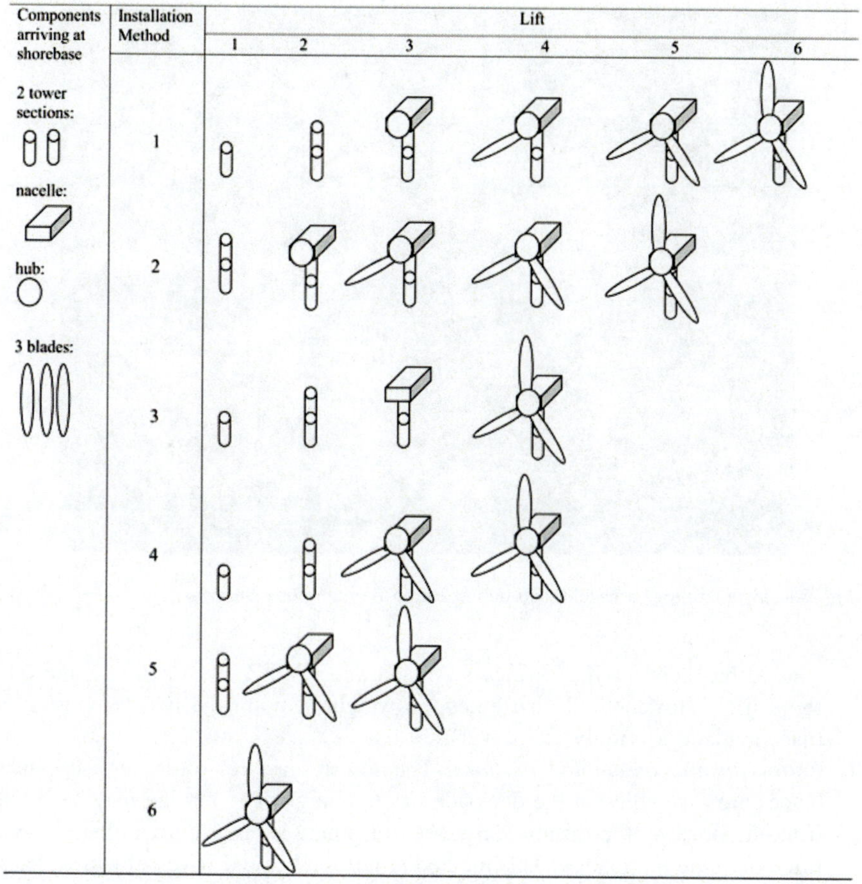

Fig. 5.5 Diagrammatic representation of installation methods

distributes the weight among the lifts and removes the need for individual blade lifts. This method requires four lifts and was employed at Nysted, Alpha Ventus (Fig. 5.7), Lillgrund, Horns Rev 2, Middelgrunden, Arklow, and Thornton Bank.

4. *Rotor and nacelle in bunny ear configuration.* The tower is transported in two pieces and lifted. The nacelle, rotor, and two of the blades are assembled onshore and lifted on site. The third blade is lifted independently. The lift capacity for the nacelle, rotor, and blade assembly is generally limiting, however, the difference between the tower weight and the nacelle assembly weight is usually small. This method requires four lifts and has been used at Horns Rev (Fig. 5.8), North Hoyle, Barrow, Scroby Sands, and Kentish Flats.
5. *Tower assembled onshore, rotor and nacelle in bunny ear configuration.* The tower is assembled onshore and installed in a single lift. The rotor is installed in the bunny ear configuration and the last blade is installed separately. This

Fig. 5.6 Installation of a nacelle and hub at Burbo Bank. *Source* Siemens

method has been used at Princess Amalia and OWEZ and requires three offshore lifts. This method distributes the weights among the two heaviest lifts (nacelle/blade assembly and tower).

6. *Entire turbine assembled onshore*. The tower, nacelle, rotor, and all three blades are assembled at the dockside or on a barge. The turbine may be lifted from the dock by the installation vessel, or otherwise transported offshore and lifted onto the foundation. This method requires a heavy-lift vessel with at least 500 tons lift capacity and was employed at Beatrice, a demonstration project (Fig. 5.9). The one-lift approach has not been employed at any other project, but various proposals have been suggested for future developments.

The method used for turbine installation determines the maximum weight lift required which in turn determines the minimum crane capacity requirement of the vessel. Table 5.1 defines the approximate maximum lift weights for vessels using the six installation methods described. The crane capacity of the existing turbine installation fleet ranges from 100 to 1200 tons. The method used to install turbines is determined by the costs of available vessels, the turbine model, and crane capacity.

5.2.3 Factors Impacting Installation

Several factors influence turbine installation time, including weather, the number of turbines installed, vessel spread, crew experience, concurrent activities, and

Fig. 5.7 Installation of an assembled rotor at Alpha Ventus. *Source* Alpha Ventus

distance to port. Turbine installation is sensitive to weather delays due to the height of lifts required and the operating constraints imposed by meteorological conditions.[1] Distance to port may also be important because there are few options for transporting components. Additionally, turbine installation times will be impacted by the degree of onshore assembly and the installation method, both of which are impacted by the turbine size and the vessel capabilities.

It is possible that new methods will reduce installation time. Installers may develop a more efficient offshore logistics system which will reduce risk associated with offshore transfers and allow the installation vessel to remain on site. Similarly, if installation vessel supply is adequate, installers may seek to develop an assembly-line approach. It has been suggested that the degree of onshore assembly may increase in the future [2], although such opportunities are currently limited by vessel deck space, crane capacity, and the sensitivity of the nacelle. Turbine installation vessels are not designed for installing fully assembled turbines.

1 Lifting operations typically require wind speeds <8 m/s and swells <0.5 m.

Fig. 5.8 Lifting of a nacelle in the bunny ear configuration. *Source* DONG

Fig. 5.9 The *Rambiz* installing a fully assembled turbine at Beatrice. *Source* REPower

5.2.4 Proposed U.S. Turbine Installation Methods

U.S. offshore wind projects have not finalized their method of turbine installation. Cape Wind developers plan to use a self-propelled, elevating installation vessel to carry the components of 6–8 turbines per trip. The tower is to be assembled in two sections, and then the nacelle, hub, and blades will be raised and secured. Pre-assembly will depend on the vessel specification and size of the deck layout.

Table 5.1 Number of offshore lifts and approximate weights for alternative turbine installation methods and turbines

		Maximum weight (ton)		
Number of lifts	Typical limiting lift	Siemens 3.6–107	Vestas V90	Repower 5M
6	Nacelle	125	70	305
5	Tower or nacelle	180–200	150	300
4	Nacelle	125	70	300
4	Nacelle/rotor assembly	190	105	380
3	Tower or nacelle/rotor assembly	180–200	150	380
1	Total assembled weight	400–420	262	655

Bluewater Wind has expressed interest in building one or more self-propelled turbine installation vessels but financing and other relevant details have yet to be worked out. Bluewater's proposed vessels are similar to those used in Europe and therefore the installation strategy may be similar as well. By contrast, Deepwater Wind plans on building a specialized heavy-lift barge and using Method 6 for installation.

5.3 Cable Installation

There are several methods for offshore wind cable[2] installation.

1. *Simultaneous lay and bury using plow*. The plow is pulled by a cable laying vessel or barge and the cable is laid by a turntable placed on the vessel. The plow buries the cable in a trench approximately 2 m deep using a high pressure water jet. The water jet fluidizes the sand or mud and the cable sinks into the trench. This is the most common method of installation, especially for export cables. This method has been used for inner-array cables at Scroby Sands and Rhyl Flats and is proposed at Cape Wind; it has been used for export cables at North Hoyle, Scroby Sands, Barrow, Rhyl Flats, OWEZ, Lynn and Inner Dowsing, and Gunfleet Sands.
2. *Simultaneous lay and bury using tracked ROV*. This method is similar to Method 1 but uses an ROV instead of a plow. The ROV carries the cable in a spool which limits its use to inner-array cables. This method has been used at North Hoyle, Barrow, and Lynn and Inner Dowsing and is the preferred method when installing cable from the turbine installation vessel.
3. *Pre-excavate*. Pre-excavate a trench using a backhoe dredge, lay cable in the trench using a cable laying vessel and fill the trench with the dredge. This method may utilize floating cables over the trench via air bags, or may lay cables directly in the trench. This has been used for inner-array and export

[2] Submarine power cable installation is documented in [3].

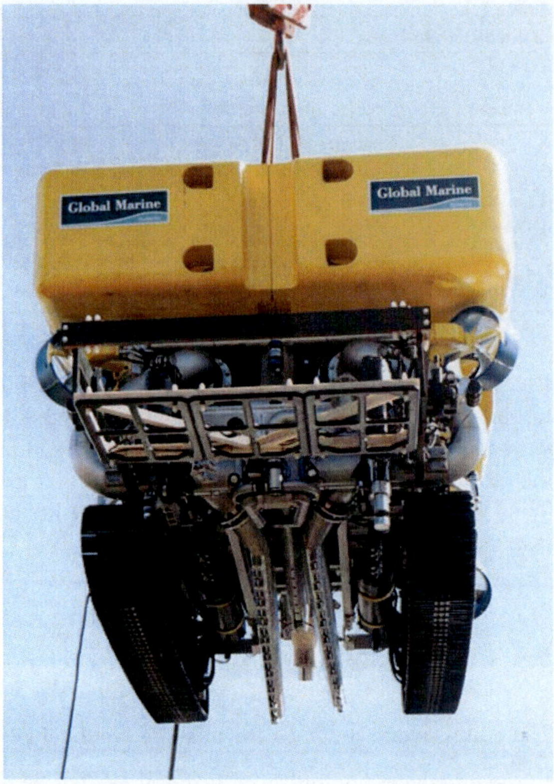

Fig. 5.10 Tracked ROV. *Source* Global Marine Power Systems

cables at Lillgrund and Middelgrunden. It was also used for a small section of export cable at Barrow.

4. *Lay and trench*. Lay cable on the seabed with a cable laying vessel and trench the cable with an ROV (Fig. 5.10). This method has been used for inner-array cable at Kentish Flats, Gunfleet Sands, and Horns Rev 2 and for export cables at Princess Amalia.
5. *Pull and trench*. Pull cables among turbines using a winch and later bury with an ROV or plow. This method is only useful for inner-array connections and was used at Horns Rev.

5.3.1 Inner-Array Cable

Connecting the inner-array cable to the wind turbines is subject to weather and operational problems and may lead to time overruns. For monopiles, a J-tube is attached to the outside to serve as a conduit for the electrical cable. The J-tube extends from above sea level down to or below the mud line. For tripods and jacket

Fig. 5.11 Cable laying plow and barge. *Source* Global Marine Power Systems

structures, the J-tube runs inside or along the foundation. The cable must be fed up through the J-tube via a winch. The process of feeding the cable usually requires divers and/or an ROV and is sensitive to tidal, wave, and current windows.

5.3.2 Export Cable

Export cables may be either high voltage (above 110 kV) or medium voltage (20–40 kV) depending on the capacity of the plant and the length of the export cable. High voltage cable is associated with offshore substations and is larger and heavier than medium voltage cable. Export cables are usually installed by a simultaneous lay and bury method because of its size and weight.

There is considerable variation in the methods used to bring cables to shore. Most frequently, export cables are brought to shore by horizontal directional drilling. A land-based drilling rig is positioned on the beach and drills directionally toward the ocean. The borehole is cased with plastic pipe and serves as a conduit for the cable to be pulled through. The cable is then fed by divers or ROVs from a cable laying vessel positioned offshore through the pipe and pulled onshore by a winch. Alternatively, a cable laying barge may be towed to shore at high tide and allow itself to be beached (Fig. 5.11). The cable laying plow is then pulled down the beach to lay cable. At the next high tide, the barge is refloated and towed out to sea continuing to lay cable toward the turbine array.

5.3.3 Factors Impacting Installation

The time to install inner-array cabling depends on the number of turbines and layout, soil type, depth of burial, and scour protection requirements. The time to install export cable depends on the location of the onshore substation, soil type, depth of burial, scour protection, and onshore transition. In shallow water, water depth does not play a significant role in installation time. Cable laying on the seafloor is expected to proceed rapidly, but landfall and connecting to an offshore substation will involve additional time.

The process of laying cable is affected by the weight and length of the cable. Inter-turbine cables are generally transported and installed in lengths approximately equal to the distance between turbines (usually less than 800 m), while export cables are at least as long as the distance to shore (3–60 km). As a result, a length of inner-array cable may weigh 10–20 tons, while an export cable may weigh 500–700 tons for a near-shore (10 km) wind farm. The sizes and weights of cable impact the vessels required and the installation time.

Burial depth is influenced by soil type, the probability of scour, and government regulations. Increasing burial depth increases the likelihood of unsuitable subsurface conditions. Scour protection may be required to ensure the necessary depth is maintained.

The method and vessels used represent tradeoffs and impact the installation time. Installation is fastest when using dynamically positioned vessels because the mooring spread does not need to be frequently repositioned. If Method 3 is used, the time that the cable laying vessel requires would be short, but the total time to install cables would be long due to the time needed to excavate and fill the trench. Likewise, an ROV operated from a turbine installation vessel might not be the most time-efficient solution, but the fact that it allows a single vessel to complete two jobs simultaneously could make it preferred over alternatives.

5.3.4 Proposed U.S. Cable Installation Methods

In Cape Wind, developers plan on installing the inner-array cable using jet plowing with a support tug and barge. Scour mats and rock armor will be used in some places and will require diver support. The export cable will make landfall via a horizontal borehole drilled from the land toward the offshore exit point. A temporary cofferdam will be utilized and backfilled after the operation is complete.

5.4 Substation Installation

Offshore substations are placed on monopile, jacket, or gravity foundations. The transformer is assembled onshore, lifted off the dock by a heavy-lift vessel, and transported to site (Fig. 5.12). After the transformer is placed on the foundation and secured, finishing work is performed.

Fig. 5.12 Installation of an offshore substation. *Source* DONG

5.5 European Installation Time Statistics

The purpose of this section is to evaluate and synthesize installation experiences of European wind farms. Our objective is to quantify and describe work activity statistics and correlations to inform and baseline future U.S. development.

5.5.1 Data Source

Table 5.2 shows the primary sources for installation data and identifies the best published descriptions. For some aspects of installation, no information, or only poor quality information, was available. For other aspects, detailed and consistent information was reported. Installation times are reported across multiple activities and were not normalized for distance to staging area, weather disruptions, and similar events. The unit time statistics include the impact of these factors and represent the total time to install the system rather than the time to install any single component.

5.5.2 Foundation

Table 5.3 summarizes information on the time required to install monopiles and transition pieces. Foundation installation, including transit time, weather delays,

Table 5.2 Sources of information on installation activities at select offshore wind farms

Wind farm	Sources
Horns Rev 2	[4]
Horns Rev 1	[5]
Middelgrunden	[6*, 7]
North Hoyle	[8*]
Alpha Ventus	[9]
Nysted	[10, 11]
Thornton Bank	[11, 12]
Scroby Sands	[11, 13*]
Kentish Flats	[14]
Lynn and Inner Dowsing	[15]
Barrow	[16*, 17]
Q7/Princess Amalia	[18]
OWEZ	[11, 19*]
Lillgrund	[20, 21*]

Note * Indicates a particularly detailed description of the installation process

and transition piece placement, takes on average 3.7 days per pile with a standard deviation SD = 2.1 days. Installation time ranged from 1.8 to 8.6 days per foundation. Excluding Arklow (a small, seven turbine project), the average time per foundation decreased to 3.3 days per monopile (SD = 1.5). In two cases, monopiles were driven by one vessel and transition pieces were installed by a second vessel. In these cases, it is the number of boat days[3] rather than the total time that is the meaningful statistic. After adjusting the total installation time for these cases, the average time increased to 4.0 boat days per foundation (SD = 2.0). Excluding Arklow, average time per foundation was 3.6 boat days per monopile (SD = 2.1).

Installation rates as a function of project size are shown in Table 5.4. For projects with 30 foundations or less, 4.6 boat days per foundation are required, decreasing to 3.8 boat days per foundation for 30–60 foundation installations, and 2.6 boat days per foundation when 60 or more foundations were installed. There are indications of scale economies, but detailed statistical analysis is not meaningful due to the small sample size.

5.5.3 Turbine

Table 5.5 depicts the time required to install turbines. The average time to install a turbine was 4.1 days (SD = 3.0). In six of the 18 cases, two vessels were employed and so activity time was normalized on a boat day basis. The average time to

[3] A boat day is one boat used for one day. Two boats used for one day would equal two boat days.

Table 5.3 Offshore wind farm installation requirements—foundations

Project	Number of vessels	Time (months)	Foundation type	Number of foundations	Install rate (days/found)
Middelgrunden	1	2	Gravity	20	3
Nysted	1	13[a]	Gravity	73	5.3
Thorton	1	1.1	Gravity	6	5.5
Lillgrund	1	14	Gravity	48	8.8
Sprogo		0.5	Gravity	7	2.1
Horns Rev 2	1	5.5	Mono	92	1.8
North Hoyle	2	4	Mono	30	4 (5.5)[b]
Rhyl Flats	1	3	Mono	25	3.6
Scroby Sands	1	3.5	Mono	30	3.5
Kentish Flats	1	2	Mono	30	2
L/I Dowsing	1	8	Mono	54	4.4
Barrow	1	7	Mono	30	7
P. Amalia	1	6	Mono	60	3
OWEZ	1	4	Mono	36	3.3
Horns Rev 1	2	4	Mono	80	1.5 (3)
Burbo Bank	1	1.8	Mono	25	2.2
Robin Rigg	1		Mono	60	
Arklow	1	2	Mono	7	8.6

Notes [a] Includes time for dredging which would not require major installation vessels
[b] Number in parenthesis is boat days per foundation. Boat days for North Hoyle is not twice the time because the two vessels did not operate for the same period of time

Table 5.4 Rate of foundation installation by number of foundations

Number of foundations	Installation rate (boat days/foundation)	Number of observations
≤ 30	4.6	7
30–60	3.8	2
≥60	2.6	3

install a turbine was 5.7 boat days (SD = 5.7). When small projects were excluded (Thornton Bank, Alpha Ventus and Arklow), the average dropped to 4.0 days per turbine (SD = 2.6). In some cases, turbines were installed in under 2 days per turbine including at Lynn and Inner Dowsing, Horns Rev 2, Nysted, Lillgrund, and Burbo Bank.

The relationship between the installation rate and method of installation is presented in Table 5.6. Methods 1 and 2 required less than 2 boat days per turbine; Methods 3, 4, and 5 were associated with longer average installation times. This trend is unexpected and may reflect the small sample size.

Table 5.7 shows the relationship between the installation rate and the total number of turbines installed. As in foundation installation, a general trend of faster installation with increasing turbine number is observed, and provides evidence of learning. Again, the sample sizes are too small and the standard deviations are too large to make any definitive statement regarding turbine number and installation time.

Table 5.5 Offshore wind farm installation requirements—turbines

Project	Number of vessels	Duration (months)	Number of turbines	Installation rate (days/turbine)	Installation method
Lillgrund	1	2.5	48	1.6	3
OWEZ	1	3.5	36	2.9	5
Kentish Flats	1	4	30	4.0	4
Scroby Sands	1	3	24	3.8	4
Nysted	1	3	72	1.3	3
Horns Rev 1	2	4	80	1.5 (3.0)	4
Burbo Bank	1	1.5	25	1.8	2
P. Amalia	2	11	60	5.5 (9.5)	5
Middelgrunden	2	1.25	20	1.9 (3.8)	3
North Hoyle	2	3	30	3.0 (6.0)	4
Alpha Ventus	1	1.5	6	7.5	3
Thornton Bank	2	2.5	6	12.5 (25.0)	3
Robin Rigg	2	9	60	4.5 (9.0)	4
Horns Rev 2	1	6	91	2.0	3
L/I Dowsing	1	3.5	54	1.9	1
Barrow	1	5	30	5.0	4
Arklow	1	2	7	8.6	3

Note Number in parenthesis is boat days per foundation. Princess Amalia boat days are not twice the total time because one vessel was not used for much of the project

Table 5.6 Rate of turbine installation by installation method

Installation method	Number of observations	Installation rate (days/foundation)
1	1	1.9
2	1	1.8
3	7	7.1
4	6	5.1
5	2	6.2
6	0	

Table 5.7 Rate of turbine installation by number of turbines

Number of turbines	Number of observations	Installation rate (days/foundation)
<10	3	13.7
10–30	6	4.1
31–60	5	5.0
>60	3	2.1

5.5.4 Cable

Cable can be laid rapidly but the connection points at turbines and pull-ins to shore are subject to weather and operational delays. Table 5.8 depicts the total time required to install export and inner-array cables and the installation time per km of

Table 5.8 Offshore wind farm installation requirements—cables

Project	Number of export cables	Length (km)		Total time (days)		Installation rate (km/day)	
		Export	Inner-array	Export	Inner-array	Export	Inner-array
Middelgrunden	2	3	14	–	30	–	0.47
Nysted	1	11	48	–	–	–	–
Horns Rev 1	1	–	57	–	90	–	0.63
North Hoyle	2	12	15	49	105	0.5	0.14
Scroby Sands	3	4	17	60	60	0.2	0.28
Barrow	1	26	22	130	150[a]	0.2	0.15
Lillgrund	1	7	22	5[b]	90	1.4	0.24
Robin Rigg	2	13	32	60[c]	180[c]	0.4	0.18
Alpha Ventus	1	60	16	–	–	–	–
Thornton Bank	2	36	51	70	161	1.0	0.32
Rhyl Flats	3	11	18	60[d]	90	0.6	0.20
Kentish Flats	3	10	21	30	60	1.0	0.35
L/I Dowsing	6	7	19	–	–	–	–
P. Amalia	1	28	45	30	150	0.9	0.30
Horns Rev 2	1	42	70	–	–	–	–
OWEZ	3	15	27	40[c]	90[c]	1.1	0.30
Burbo Bank	3	19	21	–	–	–	–
Sprogo	–	–	5	–	-	–	–

Notes [a] Cables laid by ROV while turbine was assembled; therefore, includes time for transit by turbine installation vessel.
[b] Does not include pre-excavation or delays due to ship damage.
[c] Planned times.
[d] Approximate installation times.

cable. Export cables were laid at an average rate 0.7 km/day (SD = 0.4) and inner-array cables were laid at an average rate of 0.3 km/day (SD = 0.1). Export cables rates ranged between 0.2 and 1.4 km/day; inner-array cable rates ranged from 0.1 to 0.6 km/day.

Rate estimates are conservative and uncertain. In general, reports of vessel utilization for cable laying are not as common as those for foundation and turbine installation, and so the sample size is smaller. Distance to port, burial depth,[4] and scour protection are not distinguished in the unit time statistics.

Table 5.9 shows the installation rate by total cable length for inner-array cables. The rates are fairly uniform by type and do not vary significantly with distance or within category. In Table 5.10, export cable installation time by voltage and distance is depicted. Medium voltage cable is smaller and lighter than high voltage cable and has a faster lay rate and as distance increases, so does the lay rate. This probably reflects the fixed time component of the installation of the onshore

[4] For the most part, burial depth falls within a narrow range and is probably not an important factor in the sample statistics.

Table 5.9 Inner-array cable installation rate by distance

Distance (km)	Number of observations	Installation rate (km/day)
<20	4	0.3
20–30	4	0.3
31–50	2	0.2
>50	2	0.5

Table 5.10 Export cable installation rate by voltage and distance

Length (km)	High voltage, > 132 kV (km/day)	Medium voltage, < 36 kV (km/day)	Average (km/day)
≤30	0.57	0.73	0.66
>30	0.85	1.00	0.90
Average	0.68	0.78	

transition. Export cable has a faster lay rate than inner-array cable which is likely due to the smaller number of connections that have to be made.

5.5.5 *Substation*

There is little reliable data on substation installation. The best record of the installation of a substation is from Thanet, where the jacket and substation were installed by the *Stanislaw Yudin* in four days. The Cape Wind EIS states that installation of its substation will require one month, but much of this time will be spent doing finishing work, and heavy-lift vessel support will only be required for a few days during this time. Foundation installation requires approximately the same amount of time and same vessels required for monopile or jacket installation. We expect a substation monopile foundation to take approximately four days to install and slightly longer if a jacket is used. Placement and securing the substation could be accomplished in as little as one day. However, since a slow moving heavy-lift vessel is required to travel to the site, one or more days would need to be added, depending on the distance to the staging area.

References

1. MMS (2008) Cape Wind final environmental impact statement. Minerals Management Service. Herndon, VA. 2008–040
2. Prowse M (2009) Installation logistics for UK round 3 wind farms. BWEA Offshore Conference, Westminster, June 24–25
3. Worzyk T (2009) Submarine power cables: design, installation, repair, environmental aspects. Springer, Berlin

4. Lindving K (2009) A2Sea News. www.a2sea.com/Admin/Public/DWSDownload.aspx?File=%2FFiles%2FFiler%2FNewsletters%2FA2SEA_NEWS_Spring_2009.pdf. robin rigg prowse. Accessed 29 Oct 2009
5. Elsam (2002) Horns Rev newsletter. http://www.hornsrev.dk/Engelsk/nyheder/nyh_aug_02/uk-aug_02.htm. Accessed 10 Nov 2010
6. Larsen JHM, Soerensen HC, Christiansen E, Naef S, Vølund P (2005) Experiences from Middelgrunden 40 MW offshore wind farm. Copenhagen Offshore Wind Conference. October 26–28
7. Sørensen HC, Hansen LK, Larsen JHM (2002) Middelgrunden 40 MW offshore wind farm Denmark-lessons learned. Renewable realities–offshore wind technologies. Orkney, UK
8. Carter M (2007) North Hoyle offshore wind farm: design and build. Proceedings of the Institution of Civil Engineers, Energy 160(1):21–29
9. Alpha Ventus (2010) Alpha Ventus. http://www.alpha-ventus.de/index.php?id=80. Accessed 26 Oct 2010
10. Volund P (2005) Concrete is the future for offshore foundations. Proceeding of Copenhagen Offshore Wind, October 26–28, Copenhagen
11. Gerdes G, Tiedmann A, Zeelenberg S (2006) Case study: European offshore wind farms—A survey for the analysis of the experiences and lessons learnt. Power. Deutsche Windgard, Deutsche Energie Agentur
12. C-Power (2010) Investment. http://www.c-power.be/English/investment/index.html. Accessed 26 Oct 2010
13. Douglas Westwood, ODE (2005) Scroby Sands supply chain analysis. Department of Trade and Industry, London, 334–04
14. Offshore Wind Capital Grants Scheme (2007) Kentish Flats offshore wind farm 2nd annual report. Department of Business Enterprise and Regulatory Reform, London. 09/P46. http://webarchive.nationalarchives.gov.uk/+/http://www.berr.gov.uk/files/file50164.pdf. Accessed 7 Oct 2011
15. Centrica (2010) Renewable. http://www.centrica.com/index.asp?pageid=923#project4. Accessed 10 Nov 2010
16. BoWind (2008) Construction. http://www.bowind.co.uk/construction.shtml. Accessed 10 Nov 2010
17. Offshore Wind Capital Grants Scheme (2007) Barrow offshore wind farm first annual report. Department of Business Enterprise and Regulatory Reform, London. 09/P45. http://webarchive.nationalarchives.gov.uk/+/http://www.berr.gov.uk/files/file50163.pdf. Accessed 24 Oct 2010
18. De Vries E (2007) North Sea construction: Installing the monopiles for the Dutch Q7 offshore wind farm. Renewable Energy World, May 1
19. NordzeeWind (2008) Offshore wind farm Egmond aan zee general report. Ijmuiden, Netherlands. OWEZ_R_141_20080215. http://www.noordzeewind.nl/files/Common/Data/OWEZ_R_141_20080215%20General%20Report.pdf. Accessed 24 Oct 2010
20. Vattenfall (2010) Wind farms. Stockholm. http://www.vattenfall.com/en/wind-farms.htm. Accessed 24 Oct 2010
21. Jeppsson J, Larsen PE, Larsen A (2008) Technical description of Lillgrund power plant. Vattenfall. 218258-1

Chapter 6
Installation and Vessel Spread Requirements

6.1 Vessel Categorization

6.1.1 Main Installation Vessels

Several types of turbine and foundation installation vessels exist including liftboats, jackup barges, self-propelled installation vessels (SPIVs), and heavy-lift vessels. Liftboats, jackup barges, and SPIVs are collectively referred to as elevating vessels because they elevate above the water line. SPIVs are also called turbine installation vessels (TIV) because they are used almost exclusively for these operations. We use the term turbine installation vessel to refer to any vessel capable of installing turbines or foundations and SPIV as a specific class of TIV. Basic information about the most commonly used European installation vessels is given in Table 6.1. Primary specifications for main installation vessels are deck load, leg length, and crane capacity.

6.1.1.1 Liftboats

Liftboats are self-propelled barge-shaped vessels designed with jack-up legs to create a rigid elevating platform (Fig. 6.1). Liftboats typically have three long legs which allow them to work at elevated heights with short-boomed cranes. Liftboats range in size from small vessels capable of carrying 75 tons and lifting 50 tons to much larger vessels capable of carrying 750 tons and lifting 500 tons. Large liftboats with crane capacities of 200 tons or more and deck loads of at least 500 tons are capable of carrying one to two turbines. Small liftboats are not capable of most offshore wind installation work.

M. J. Kaiser and B. F. Snyder, *Offshore Wind Energy Cost Modeling*,
Green Energy and Technology, DOI: 10.1007/978-1-4471-2488-7_6,

Table 6.1 Vessels used in offshore wind farm construction in Europe

Vessel	Vessel type	Operational water depth (m)	Crane capacity (ton)	Wind farms
Sea Power	SPIV	24	100	Horns Rev 1, Lillgrund, Horns Rev 2
Sea Energy	SPIV	24	100	Kentish Flats, Scroby Sands, Nysted, Princess Amalia
Rambiz	Shearleg crane	>100	3,300	Beatrice, Thornton Bank, Nysted
Sea Jack	Jackup barge	35	1,300	Princess Amalia, Arklow, Scroby Sands
Svanen	Heavy-lift vessel	>100	8,700	OWEZ, Rhyl Flats, Gunfleet Sands
Titan 2	Liftboat	60	400	Rhyl Flats
Buzzard	Jackup barge	45	750	Alpha Ventus, Thornton Bank
JB 114 and 115	Jackup barge	50	280	Alpha Ventus
Thailf	Heavy-lift vessel	>100	14,200	Alpha Ventus
Eide Barge 5	Shearleg crane	>100	2,000	Middelgrunden, Nysted, Lillgrund, Sprogo
Taklift 4	Shearleg crane	>100	1,600	Alpha Ventus
Kraken and Leviathan	SPIV	40	300	Walney, Greater Gabbard
Resolution	SPIV	35	300	Robin Rigg, Barrow, Kentish Flats, North Hoyle
Excalibur	Jackup barge	30	220	North Hoyle,
Lisa A	Jackup barge	50	600	Rhyl Flats
MEB JB1	Jackup barge	40	270	Middelgrunden, North Hoyle, Yttre Stegrund
Goliath	Jackup barge	50	1,200	
Sea Worker	Jackup barge	40	400	Robin Rigg; Gunfleet Sands

Fig. 6.1 The KS *Titan II* liftboat. *Source* Offshore Wind Power Marine Services

Fig. 6.2 The *Sea Jack* jackup barge. *Source* Ben Barden/Vattenfall

6.1.1.2 Jackup Barges

Jackup barges typically have four legs and are usually intermediate in size between liftboats and SPIVs. Figure 6.2 shows a large jackup barge (the *Sea Jack*) while Fig. 6.3 shows a smaller jackup barge (the *JB 114*). The *Sea Jack* has a crane capacity of 800 tons and a deck load of 1500–2000 tons while the *JB 114* has a crane capacity of 280 tons and a deck load of 1250 tons. A small jackup barge may be able to carry two turbines while a large jackup might carry six to eight turbines. Jackup barges are not self-propelled and require a tow to site; transit speed depends on tug power and normally ranges between 4 and 8 knots.

Fig. 6.3 The *JB 114* jackup barge at Alpha Ventus. *Source* Alpha Ventus

6.1.1.3 SPIVs

Self-propelled installation vessels are large vessels with four to six legs that travel at 8 to 12 knots and have variable deck loads of 1,500 to 6,500 tons (Fig. 6.4). Most SPIVs are elevating and ship-shaped, but they may also be column stabilized (rather than fully elevating) or barge shaped. They are distinguished from jackup barges by propulsion and from liftboats by size. Depending on the deck load or deck space, they usually carry six to eight turbines.

6.1.1.4 Heavy-lift Vessels

Heavy-lift vessels are barge-shaped hulls with high capacity cranes that do not employ an elevating system; they may or may not be self-propelled and may be dynamically positioned or conventionally moored. Heavy-lift vessels include shearleg cranes, derrick barges, and other floating cranes and are widely used in

Fig. 6.4 The SPIV *MPI Resolution. Source* Chris Laurens/Vattenfall

offshore oil and gas construction. They are rarely used to install turbines, but may be used for foundation work, installing fully assembled turbines, or installing substations (Fig. 6.5). Heavy-lift vessels travel at 4 to 8 knots.

6.1.2 Cable-Laying Vessels

Cable-laying vessels are used to lay the inner-array cable between turbines and between the export cable to shore. Export cable laying vessels are large barges or self-propelled vessels dedicated specifically for cable laying operations (Fig. 6.6). These vessels typically have a turntable capable of spooling over 1,000 tons of cable and operate a cable laying plow or ROV. Inner-array cables may be laid by a variety of vessels because cable distance and weight restrictions are smaller. Barges, modified offshore supply vessels, or other medium-sized vessels are commonly used and may be dynamically positioned and operate a cable laying ROV.[1] It is also common for the main installation vessel or the export cable vessel to be employed in inner cable laying.

6.1.3 Vessel Spreads

Support vessels are required for all installation stages but the size and composition of the support spread will vary over the course of installation and depends on the capability of the main installation vessel and the scope of work. There are several

[1] The ROV may carry a spool of cable itself or may be supplied with cable by the installation vessel.

Fig. 6.5 The heavy-lift vessel *Rambiz* installing a foundation at Ormonde. *Source* Ben Barden/ Vattenfall

Fig. 6.6 The *CS Sovereign* cable laying vessel. *Source* Global Marine

Fig. 6.7 Spread vessels; *From Top* crewboat, anchor handling tug, multicat. *Source* Offshore Wind Power Marine Services; Darren Hillman; Albert Bridge

classes of spread vessels used in offshore wind installations including crewboats, multicats, tugs, dive support vessels, and dredging/scour vessels (Fig. 6.7).

Crewboats are 10–25 m long and carry between 10–15 people. They range from small rigid hulled inflatable boats to larger catamarans. In addition to their

Table 6.2 Vessel capabilities by work type

	Installation Activity			
Vessel class	Foundation	Turbine	Cable	Substation
Liftboat	Unlikely	Yes	No	No
Jackup barge	Yes	Yes	No	Yes
SPIV	Yes	Yes	Yes	Yes
Heavy-lift	Yes	Unlikely	No	Yes

personnel transfer role, these vessels may be used as utility vessels for energizing turbines, enforcing safety zones, conducting environmental studies, or supporting shallow water divers.

Multicats are multipurpose vessels, typically 12–30 m in length and usually equipped with a small (under 50 ton) crane and a large open deck. They are primarily used for anchor handling and may be used for light transport duties, diver support, dredging, and other tasks. Multicats are not common in the U.S.

Dive support can be conducted from a variety of vessels with minor modifications. Dive support vessels provide a place to launch, supply, communicate with, and recover divers; they are much smaller than vessels used in the offshore oil and gas industry since operations are in shallow water and saturation diving is not necessary.

Dredging and scour protection vessels vary with the application. In their simplest form, a dredge may consist of a backhoe excavator placed on a spud barge. More sophisticated dynamically positioned trailing suction hopper dredges are used for gravity foundations but are not required for most monopile development. Scour protection is typically placed by a side dumping barge but could also be placed by a utility vessel.

Tugs are used to tow deck barges and non-self-propelled vessels from the shore base to the offshore site. In many cases they are equipped with a small crane for anchor handling. Tugs vary in power, size, and ocean-going capability and are specified by horsepower. Vessels used in offshore wind support usually range from 1,000 to 3,500 kW (approximately 1,500 to 5,000 hp).

6.2 Factors Impacting Vessel Selection

Vessels are matched with projects based on availability, economic, and technical factors. Technical demands depend on the stage of installation. Table 6.2 shows the vessel types capable of contributing in each stage of installation. The operational water depth is critical for all installation stages. Both deep and shallow water can be limiting; in shallow areas, jackup barges may be required due to their lower draft [1]. To date, deep water has not been limiting due to the preference for shallow water sites close to shore.

6.2.1 Foundation

Foundations may be installed by any elevating installation vessel, but are not likely to be installed by liftboats due to their low crane capacity. Foundations may be installed by floating heavy-lift vessels but for monopile foundations, elevating vessels have generally been used. Important factors in the specification of foundation installation vessels are crane capacity and water depth. Crane capacity is important because monopiles can weigh over 500 tons and transition pieces can weigh over 200 tons which can exceed the lift capacity of many cranes. If a vessel is unable to lift a monopile, alternative methods for handling may be used, but no alternative is readily available for handling the transition piece. Crane lift height is usually not an important factor because foundations only need to clear the vessel deck. Transit speed is also not critical because there are alternative methods for foundation transport that do not require the installation vessel to move back and forth from port.

6.2.2 Turbine

Turbines may be installed by any elevating installation vessel and in order to work on a particular project, a vessel must be capable of carrying a suitable number of turbine components and lifting these components to hub height. Critical factors in vessel selection include variable load, deck space, and maximum lift height and weight. Variable load interacts with deck space to determine the weight, number, and assembly state of turbine components carried per trip. Lift height determines whether a vessel can install a turbine at a given hub height and is a function of water depth, leg length, and crane boom length (Fig. 6.8). The combination of air gap and boom length must reach above hub height in order to install the nacelle and blades. Air gap is determined by leg length, water depth, and weather conditions. Crane lift capacity determines the number of lifts required per turbine and sets limits on the degree of onshore assembly allowed. Onshore assembly reduces the number of offshore lifts and consequently the time and cost of the operation, but as the degree of onshore assembly increases, the crane capacity and deck space required increases.

6.2.3 Cable

For both inner-array and export cable, the vessel equipment available determines the installation technique; in most cases, the installation vessel operates a plow or ROV that simultaneously lays and buries the cable, and this is the most time-efficient option, but depending on vessel availability and cost, may not always be chosen. For export cable, turntable capacity is critical. High voltage export cables

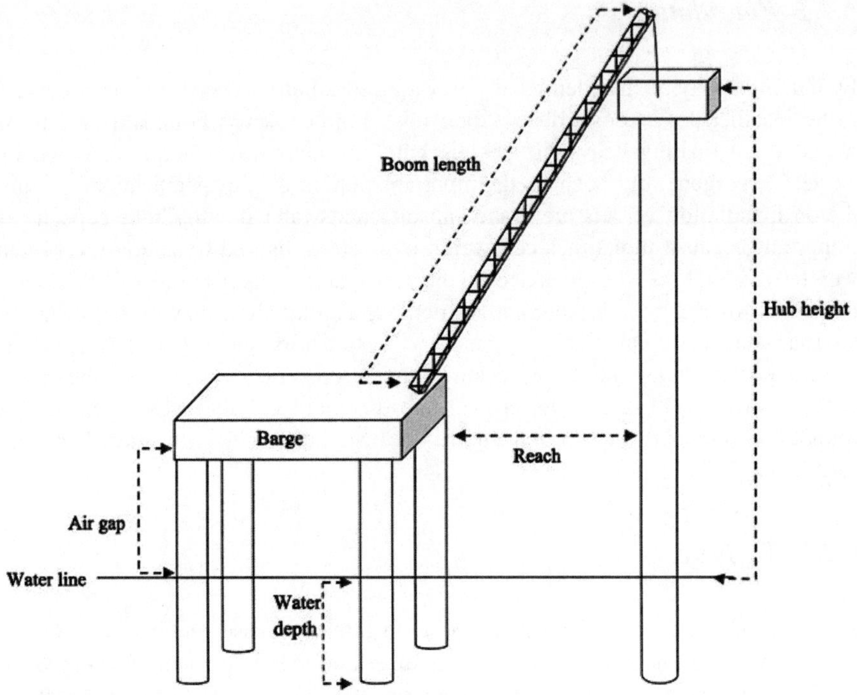

Fig. 6.8 Diagram of factors affecting lift height and capacity

are typically very long and heavy and require a vessel capable of feeding hundreds or thousands of tons of cable. For inner-array cable, turntable capacity is less critical. Inner-array cables are transported and installed in smaller lengths (under 1 km) and are lighter than export cables.

For inner-array cables, self-propelled vessels are typical. For export cables, barges or self-propelled vessels may be used. A barge simplifies the onshore transition since it can be beached, but has the disadvantage of requiring frequent repositioning of its mooring spread.

6.2.4 Substation

Both substation jackets and topsides typically weigh 500–2,000 tons and the installation vessel must have the crane capacity to lift the substation. Lift height is usually not critical as substations are generally not tall, however, if topsides need to be elevated significantly due to potential wave heights, lift height could be a

Table 6.3 Estimated potential spreads by vessel type and activity

			Number of tugs		Number of barges		Number of utility and crew vessels	
Vessel type	Component	Transport system	Min	Max	Min	Max	Min	Max
SPIV	Foundations	Barge/float	2	3	0	3	2	4
SPIV	Foundations	Self-carry	0	0	0	0	1	3
SPIV	Turbines	Self-carry	0	0	0	0	1	3
Jackup	Foundations	Barge/float	3	4	0	3	2	4
Jackup	Foundations	Self-carry	1	2	0	0	1	3
Jackup	Turbines	Self-carry	1	2	0	0	1	3
Liftboat	Turbines	Self-carry	0	0	0	0	1	3

factor, especially for shearleg cranes. To date, substation jackets and topsides have been installed by heavy-lift vessels; as crane capacity of jackup barges and SPIVs increases, they may be used to install substations.

6.3 Support Spread

Spreads provide two major functions during installation: they support the activity of installation vessels by crew transport, anchor handling, towing and other functions, and they transport components, especially foundations, from the shorebase to the offshore site. Spread composition is likely to be highly variable and depend on installation vessel specifications, transport strategy, distance to the shorebase, expected environmental conditions, turbine and foundation weights, project size, and vessel costs and availability.

Several potential spreads required to support main installation vessels during turbine and foundation installation are shown in Table 6.3. In all cases, jackup barges will require one tug for propulsion, while SPIVs and liftboats will not. Jackup barges will also require some vessels for anchor handling. It is possible that liftboats or SPIVs would need anchor handling, but this is considered unlikely, especially if newbuilt vessels are used.

6.3.1 Foundation Transport

Foundations may be carried by the installation vessel, barged to the site, or floated. If the installation vessel carries the foundations, no additional spread vessels are needed. If the foundations are barged, at least two tugs and two barges will be needed to ensure a constant supply of foundations during construction. If the foundations are floated, at least two tugs would be required. While monopiles may be floated, transition pieces are not; they could be carried by the installation vessel or barged. If the offshore site is particularly distant from the onshore staging area, additional vessels may be required.

6.3.2 Turbine Transport

Turbine components are most likely to be carried by the installation vessel and would not require additional vessel support. If components are carried by a feeder vessel, it is likely that such a vessel would be self-elevating and/or dynamically positioned. A barge pulled by a tug is unlikely to allow for safe turbine transfers at sea. Therefore, a turbine feeder system would require the addition of another main installation vessel rather than the addition of spread vessels.

6.3.3 Cable

Cable laying vessels also require spreads. Vessels that are not self-propelled requires tugs for propulsion and anchor handling; due to the frequency of mooring positioning and the number of mooring lines, at least two tugs are utilized (Fig. 6.9). Both self-propelled and non-self-propelled vessels require at least one general purpose crew/utility vessel and one dive support vessel or other support vessel.

6.3.4 European Spread Requirements

Figures 6.10 through 6.12 show the number of vessels on site by week for the Thanet, Gunfleet Sands, and Ormonde developments. In each case, weekly notices to stakeholders were released detailing the activities and vessels on site. Empty spaces represent missing data, not inactivity. All three wind farms are UK commercial-sized projects (300, 172, and 150 MW for Thanet, Gunfleet, and Ormonde, respectively) developed in the 2009–2011 period.

Gunfleet and Thanet are built on monopile foundations, while Ormonde uses jacket foundations. Gunfleet Sands is 7 km from shore and developed by DONG; Thanet and Ormonde are 10–12 km offshore and developed by Vattenfall. Thanet contains 100, 3 MW turbines, Gunfleet Sands contains 48, 3.6 MW turbines, and Ormonde uses 30, 5 MW machines. The characteristics of the wind farm and installation strategies determine the shape and duration of each graph.

The minimum spread at Thanet is one main installation vessel (the jackup barge *Sea Jack*), two crewboats, two tugs, and one other vessel. When a second installation vessel was added (the self-propelled *Resolution*) the spread increased to four tugs, two crewboats, and one guard vessel. As cable laying began in week 27, the number of multicats and crewboats increased. The maximum number of vessels operating at any one time was 32 when two jack-ups, three cable lay vessels, and one heavy-lift ship were all on-site.

At Gunfleet Sands a larger spread was employed and the minimum number of vessels operating at one time was twelve. Typically, the spread consists of one

Fig. 6.9 Cable laying barge with spread. *Source*: Global Marine Systems

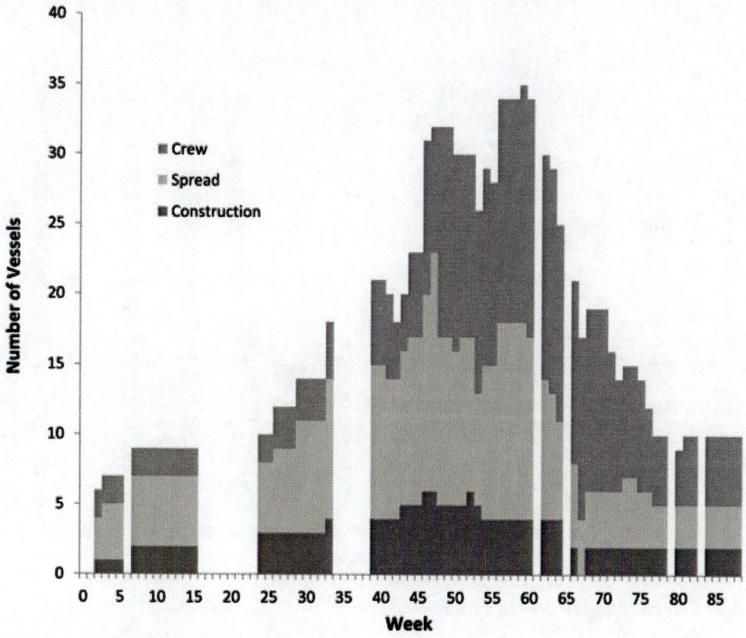

Fig. 6.10 Spread requirements at Thanet

self-propelled main installation vessel (the liftboat *Titan* 2), one tug, one barge, one cable laying catamaran, two to three multicats with at least one supporting the cable laying and one supporting the installation vessel, approximately 10 crew-boats and at least 3 other vessels.

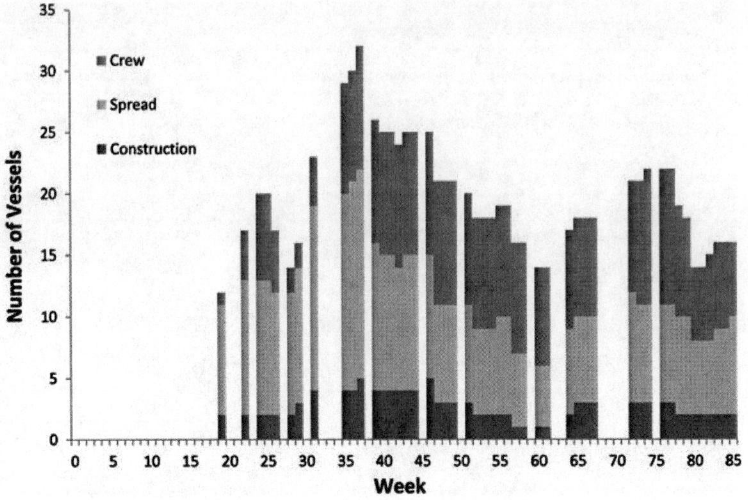

Fig. 6.11 Spread requirements at Gunfleet Sands

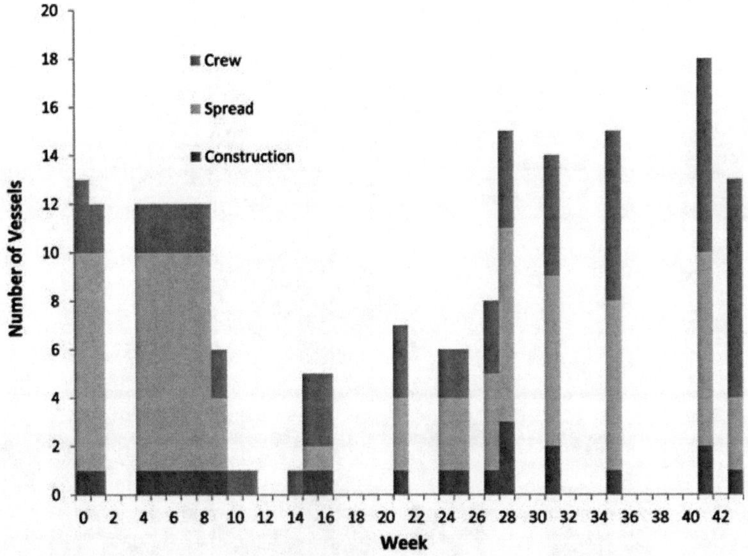

Fig. 6.12 Spread requirements at Ormonde

Ormonde uses fewer turbines than Thanet and Gunfleet Sands but the 5 MW machines are both larger and heavier requiring a different spread makeup. During foundation installation, the heavy-lift vessel *Rambiz* installed foundations with support from five tugs, four barges, and two crewboats. Cable laying was performed from a single installation barge, with one to three tugs and two to three

Table 6.4 Average number of support vessels required per construction vessel at Thanet, Gunfleet Sands, and Ormonde

Vessel type	Thanet	Gunfleet Sands	Ormonde
Crew	2.5	3.4	3
Tug	1	0.9	3.4
Other	1.4	2.7	2
Total	4.9	7	8.4

crewboats. Turbine installation required two tugs and approximately two crewboats, however, additional crewboats were added as commissioning began and personnel movements between turbines increased.

The average number of support vessels per main installation and cable laying vessel at Thanet, Gunfleet Sands, and Ormonde was 7.9,7, and 8.4, respectively. These data are further classified by support vessel type in Table 6.4. The data show significant variation in the total number of vessels required for support as well as variance in the types of vessels used, but in general, two to four crewboats are needed per installation vessel and this composes approximately half of the total spread.

6.3.5 Potential U.S. Spreads

Utilization of barges, tugs, and utility vessels in the U.S. will vary from Europe depending on availability and fleet makeup. Multicats are unlikely to be used in the U.S. because they are not common. Instead, small utility boats similar to those used in the Gulf of Mexico (GOM) oil and gas industry are likely to be employed for anchor handling. In Europe, crewboats are smaller than those used in the GOM oil and gas industry due to the need to frequently transport a small number of personnel around the site. Small crewboats from the GOM could be used, but it is likely that smaller vessels would be converted for personnel transfer or specialized newbuilds constructed in local shipyards. Such specialized vessels may allow for safer crew transfers and may be less expensive than the GOM-sized vessels.

There are a large number of dynamically positioned offshore supply vessels in the GOM which could be mobilized to wind farm sites. These vessels could be used to shuttle wind farm components from shorebases to offshore wind facilities.[2] Liftboats could also be used for turbine installation, and could eliminate the need for some barges and tugs.

---

[2] For example, the Hornbeck 250 class OSVs are dynamically positioned, have 185 x 45 ft decks, cruise at 10 kn and can carry over 2000 tons of deck load. The typical duty of these vessels involves dynamically positioning next to a stationary rig or platform and offloading cargo. This is similar to what would be required to supply turbine installation vessels.

6.4 U.S. Vessel Procurement

Main installation vessels may be procured for the U.S. offshore wind market by moving vessels from other U.S. industries or newbuilding. Vessels could also be mobilized from Europe, but high levels of European activity and the requirements of the Jones Act will constrain use of the European fleet in U.S. development.

6.4.1 Jones Act

The Jones Act (the Merchant Marine Act of 1920) requires that all commerce between two U.S. ports be carried on a U.S. flagged, crewed, and owned vessel. The definition of a port includes any area on the Outer Continental Shelf (OCS) and it is generally thought that the Jones Act restricts developers to using U.S. vessels [2–4]. The Jones Act is enforced by the Customs and Border Protection (CBP) Agency and the Maritime Administration (MARAD) and several activities are excluded via administrative policy developed by CBP. For a discussion of the Jones Act and offshore wind in the U.S., see for example [5].

6.4.1.1 Customs and Border Protection

CBP is tasked with determining when vessels are engaged in coastwise transport and subject to the provisions of the Jones Act. For the past several decades, CBP has promulgated a liberal definition of "vessel equipment" which is not subject to the provisions of the Jones Act. In general, the CBP has allowed non-U.S. vessels to carry merchandise (typically pipelines, jumpers, risers, and umbilicals) from a U.S. port to a location on the OCS as long as the same vessel installs the merchandise. CBP has allowed non-U.S. liftboats to transport and install equipment including decks, generators, jackets, boat landings, and related equipment from U.S. ports to locations on the OCS. In these cases, the CBP has ruled that the material being carried is not merchandise but vessel equipment as it is required for the vessel to perform its intended "mission".

In July 2009, CBP issued a decision restricting the definition of vessel equipment. According to CBP, the definition of equipment as codified in the Tariff Act of 1930 is "portable articles necessary and appropriate for the navigation, operation or maintenance of the vessel and for the comfort and safety of the persons on board." This definition does not reference the mission of the vessel. Material such as pipelines, wellheads, or platform decks were deemed unnecessary for the navigation, operation, or maintenance of the vessel and CBP's interpretation of the Jones Act was deemed to be contrary to the legislative intent. As a result, CBP revoked numerous rulings allowing foreign pipelaying vessels and liftboats to operate in coastwise trade [6]. In October 2009, the CBP withdrew its July decision revoking its previous rulemaking citing a significant public response on both sides of the issue and indicated that new policy would be announced in the future [7].

In separate rulings independent of the definition of vessel equipment, the CBP allows non-Jones Act qualified vessels to lift and install jackets and other structures on the OCS under limited circumstances: the foreign vessel must proceed directly to the installation site, may not discharge passengers, and the crane, not the vessel itself, must move all material. These rulings were used to allow a foreign vessel to install an offshore meteorological tower at an offshore wind farm. The CBP also allows the transport of material from a U.S. port to an offshore vessel by non-Jones Act qualified vessels as long the destination vessel is not fixed to the seabed (i.e. not elevating) [8].

6.4.1.2 MARAD

Independently, MARAD manages exemptions to the Jones Act. These are separate from CBP's rulemaking and involve vessels that are deemed to be engaged in coastwise transport. There are two classes of vessels that may be granted exemptions: small passenger vessels and barges used in the offshore petroleum industry.[3] Small passenger vessels capable of carrying up to twelve people and over 3 years old may be granted exemptions. Vessels must be greater than 5 net ton (approximately 8 m), must be owned by a U.S. citizen, and may not carry cargo. The launch barge program allows for the use of foreign vessels to transport platform jackets offshore if no suitable U.S. vessel is available. The program applies to any vessel capable of loading, transporting and launching, or installing a jacket.

6.4.1.3 Relevance to Offshore Wind

Under existing rulemaking, the Jones Act may not apply to turbine or foundation installation. Turbine and foundation installation is similar to a liftboat carrying and installing a deck or an offshore jacket and cable installation is similar to pipelaying, both of which may be conducted by non-Jones Act qualified vessels under the current interpretation of vessel equipment. Under current rulings, the only vessels that would need to comply with the Jones Act would likely be barges and tugs. However, CBPs interpretation of what constitutes vessel equipment is under review and it is possible that the definition will be tightened to exclude foreign turbine installation vessels or cable installation vessels from transporting wind farm components. While a foreign vessel could be used to install foundations and turbines, there may be limitations on the ways in which vessels could be used and this may make the use of foreign vessels unattractive.

[3] Additionally, exemptions may be granted for national defense or in emergencies.

Table 6.5 Elevating vessels active in the U.S. and/or with U.S. flags capable of offshore wind farm construction

Name	Owner	Vessel type	Crane (ton)	Depth capacity (m)	Max hook height (ft)	Deck load (ton)	2009 dayrate (1000 $/day)
Superior Influence	Superior	Liftboat	200	60	120	750	30
Superior Respect	Superior	Liftboat	200	60	120	750	30
Superior Storm	Superior	Liftboat	250	55	100	500	30
Superior Gale	Superior	Liftboat	250	55	100	500	30
Superior Champion	Superior	Liftboat	200	55	100	500	30
Jacob	CS Liftboats	Liftboat	200	60	130	450	38
Mammoth Elevator	EBI	Liftboat	400	50	150	300	
Karlissa A	Titan	Jackup barge	300	50	180	1000	25
Karlissa B	Titan	Jackup barge	300	50	180	1000	25

Source Personal communication with company personnel

6.4.2 U.S. Fleet Circa 2011

U.S. vessels capable of contributing to the construction of offshore wind farms are depicted in Tables 6.5 and 6.6. We assume that an elevating vessel must have a crane capacity of at least 200 tons to be potentially useful.[4] Non-elevating crane vessels with a lift capacity greater than 500 tons are potentially useful for installing transformer stations and foundations. We do not consider conversions of existing vessels, nor do we consider derrick barge/pipelay vessels or semisubmersible vessels.[5] Most of the vessels identified service the GOM oil and gas industry, although some are used for marine salvage or in civil construction.

U.S. heavy-lift vessel capacity appears adequate for near-term development plans, but there are a relatively small number of elevating vessels capable of installing turbines and those that do exist are not particularly well suited to this task. Large liftboats such as the *Superior Storm* or *Superior Respect* are potentially capable of lifting rotors and nacelles to 70 m, especially in shallow water. However, they only have 500–750 ton variable deck loads and would have difficulty carrying more than one or two 3–3.6 MW turbines at a time. Therefore, alternative turbine transport strategies may be required. Dayrates shown in the tables are average spot rates and do not include spreads. Contract terms, durations and day rates may be different in the offshore wind sector, and developers will have to compete with the oil industry for access to these vessels.

---

[4] Vessels with crane capacities under 200 tons would limit the type of offshore turbine used to less than 3 MW.

[5] We do not consider such vessels because their capabilities and cost exceed those required for wind applications, although similar vessels have been used in Europe on a limited basis.

Table 6.6 Non-elevating vessels active in the U.S. or with a U.S. flag capable of offshore wind farm construction

Name	Owner	Vessel type	Crane capacity (ton)	Max hook height (ft)	2009 dayrate (1000 $/day)	Location
Lili Bisso	Bisso Marine	Shear leg	600	200		GOM
Cappy Bisso	Bisso Marine	Shear leg	700	155		GOM
Big T	T&T Marine	Shear leg	600	150		GOM
Mr 2 Hooks	Laredo	Shear leg	800	110	50	GOM
Illuminator	Laredo	Shear leg	513	130	50	GOM
IOS 800	International	Shear leg	800	175	75-100	GOM
Chesapeake	DonJon	Shear leg	1000	230		Atlantic
Left Coast Lifter	American Bridge-Flour	Shear leg	1700	328	NA[a]	Pacific
Taklift 1	Smit	Shear leg	800	260		GOM
Titan II	Global Industries	Crane vessel	880	206		GOM
William Kallop	Offshore Specialty	DB[b]	1750	272	150	GOM
Superior Performance	Superior	DB	880	170	80	GOM
Raeford	Offshore Specialty	DB	700	205		GOM
Swing Thompson	Offshore Specialty	DB	1320	206	150	GOM
Superior Pride	Superior	DB	880	170	67.5	GOM
DB 16	J Ray McDermott	DB	600	231		GOM
DB 50	J Ray McDermott	DB	4400	262	350–500	GOM
Atlantic Horizon	Caldive	DB	500	205		GOM
Pacific Horizon	Caldive	DB	700	213		GOM
DB General	General Const.	DB	700	200	96	Pacific
EP Paup	Manson	DB	1000	210	139	GOM
Wotan	Manson	DB	500		115	Pacific
Arapaho	Tetra	DB	800	200	150	GOM
DB 1	Tetra	DB	615	200	150	GOM

Notes [a] Left Coast Lifter building costs were approximately $50 million and has not been used on a dayrate chartered contract
[b] DB denotes derrick barge
Source Personal communication with company personnel

Table 6.7 Specification of newbuild designs

Design	Water depth (m)	Crane capacity (ton)	Number built	Variable load (ton)	Self propelled speed (kn)
NG 9000	45	800	3	6,500	12
SEA 2000	40	1,200	4	1,600	NA
SEA 2750	45		1	3,000	NA
SEA 3250	45		1	3,800	NA
NG 7500	40	1,000	2	6,000	12
NG 2500	52	300	2	1,300	8
NG 5300	45	500	1	2,600	8
Semco 280	50	400	1	500	–

6.4.3 Newbuilding and Modification

Table 6.7 shows the specifications of the most popular newbuild designs for the European market. All newbuilding for the wind industry is currently composed of elevating vessels. Among the most frequent European newbuild designs are the SEA 2000 and the NG 9000. These two vessels differ in their capabilities and represent two alternative methods for turbine installation; the SEA 2000 is a relatively small, inexpensive jackup barge while the NG 9000 is a large, expensive SPIV.

The SEA 2000 (Fig. 6.3) has a 55.5 by 32.2 m hull with a variable load of 1,600 tons. In 2009 a newbuilt SEA 2000 cost about $63 million. The NG 9000 is a 130.8 by 39 m vessel and is self-propelled, dynamically positioned and capable of 12 knots. It is equipped with an 800 ton crane, has a variable load of 6,500 tons, and costs about $160 million to build. Either vessel would be capable of installing turbines and monopiles, but would differ in spread requirements and work durations.

There are several shipyards in the U.S. capable of building self-propelled turbine installation vessels and large turbine installation barges as shown in Table 6.8. Shipyards were included in Table 6.8 if they delivered a large OSV, a large, deep draft vessel (excluding naval vessels), a liftboat, or an oceangoing barge in the past 2 years [9].

Modification of existing vessels is also possible. Both the *Sea Energy* and *Sea Power*, two highly utilized European installation vessels, were converted from standard cargo vessels with the addition of a leg-stabilization system and crane. Additionally, dynamically positioned vessels have proven capable of efficiently installing monopiles and transition pieces and large OSVs or other dynamically positioned vessels from the oil and gas industry could be modified for foundation work.

Table 6.8 U.S. commercial shipbuilders capable of building turbine installation vessels

Builder	Typical vessels	Location
Aker Philadelphia	Product carrier	Philadelphia PA
Atlantic Marine	OSV	AL, FL
Bay Shipbuilding	Barge	Sturgeon Bay WI
Signal (Bender)	Rig, OSV	MS, TX, AL
Boconco	Liftboat	Bayou La Batre AL
Bollinger	Barge, OSV	LA
Candies Shipbuilding	OSV	Houma LA
Conrad Industries	Liftboat	Morgan City LA
Corn Island Shipyard	Barge	Lamar IN
Dakota Creek Industries	OSV	Anacortes WA
Eastern Shipbuilding	OSV, Barge	Panama City FL
Elevating Boats, Inc.	Liftboat	Laffite LA
GD NASSCO	Product carrier	San Diego CA
Gunderson Marine	Barge	Portland OR
Halimar Shipyard	Liftboat, OSV	Morgan City LA
Keppel AMFELS	Rig	Brownsville TX
Le Tourneau	Rig	Vicksburg MS
Leevac Industries	OSV	Jennings LA
North American Sbldrs.	OSV	Larose LA
Quality Shipyard	OSV	Houma LA
Rodriguez Boatbuilders	Liftboat	Bayou La Batre AL
SENESCO	Barge	North Kingstown RI
Thoma-Sea Shipbuilders	OSV	Lockport LA
US Barge	Barge	Portland OR
VT Halter Marine	Barge, OSV	MS
Zidell Marine	Barge	Portland OR
Semco	Liftboat	Laffite LA

Source [9]

References

1. Fan H, Lin J, Shi Q (2010) Installation vessel and method for offshore wind turbine in ultra-shallow water. ASME 29th Conference on Ocean, Offshore and Arctic Engineering, 6–11 June Shanghai, China
2. Graham J (2009) U.S. offshore wind project updates. Renewable Energy World. http://www.renewableenergyworld.com/rea/news/article/2009/12/us-may-see-offshore-wind-project-in-water-by-2012. 16 Dec 2009
3. Trabish HK (2010) The emerging opportunity in offshore wind vessels. Greentech Media. http://www.greentechmedia.com/articles/read/the-emerging-opportunity-in-offshore-wind-vessels. 30 Dec 2010
4. Griffin R (2010) 90 year old law may hinder offshore wind: DOE. Platts Energy Week. http://www.plattsenergyweektv.com/story.aspx?catid=293&storyid=114211. Accessed 4 Oct 2010
5. Papavizas CG, Morrissey GA (2010) Does the Jones Act apply to offshore alternative energy projects. Tulane Marit Law J 34:378–441

6. Ressin C (2009) Proposed modification and revocation of ruling letters relating to the Customs position on the application of the Jones Act to the transportation of certain merchandise and equipment between coastwise points. U.S. Customs and Border Protection, Washington, DC. CBP Bulletins and Decisions. 17 July 2009
7. Quillen K (2010) Stakes are high for Louisiana maritime industry in debate over decades-old law. The Times-Picayune. New Orleans, LA, 31 Jan 2010
8. CBP (2010) Customs ruling online search system. http://rulings.cbp.gov/. Accessed 29 Oct 2010
9. Colton T (2010) Shipbuilding history. http://shipbuildinghistory.com/. Accessed 29 Oct 2010

Chapter 7
Modeling Offshore Turbine Installation Vessel Dayrates in the U.S.

7.1 European Market Data

Anecdotal information on European installation vessels serve as a comparison for the U.S. market, but these rates need to be used cautiously because they vary with time and regional conditions, and are unlikely to extrapolate directly to U.S. markets. In 2009, SeaJacks signed contracts for the installation of wind turbines at Walney and Greater Gabbard in the United Kingdom for 148,000 and 176,000 $/day excluding mobilization costs.[1] Both contracts were for approximately 15 months duration and used identical newbuilt self-propelled vessels (*Kraken* and *Leviathan*). In late 2008 and early 2009, Master Marine signed contracts for the newbuilt *Service Jack* and *Service Jack 2*. The *Service Jack* was contracted for three years with ConocoPhillips (serving as an accommodation platform); *Service Jack 2* was let for nine months installing turbines at Sheringham Shoal, UK. The effective dayrates were $330,000 and $380,000 per day. Both vessels are large (330 ft long and 150 ft wide) and capable of operating in deep water (300 ft). Elevating installation vessel dayrates typically range from $50,000 to $250,000 per day depending on the capabilities of the vessel [1, 2].

7.2 Dayrates as a Proportion of Newbuild Costs

Dayrates can be estimated from the capital costs of construction [3]. There are no available data on the relationship between dayrates and capital costs of offshore wind installation vessels, however, this information can be computed for jackup drilling rigs, liftboats, and offshore supply vessels to provide an analog for offshore wind.

---

[1] Effective dayrates were computed by dividing the value of the contract by the number of days of operation.

M. J. Kaiser and B. F. Snyder, *Offshore Wind Energy Cost Modeling*,
Green Energy and Technology, DOI: 10.1007/978-1-4471-2488-7_7,

We obtained data from RigLogix [4] on the dayrates and construction costs of jackup rigs delivered between 2000 and 2010. All rigs for which data was available were included and a total of 102 rigs comprised the sample. We averaged all dayrate contracts for each rig over the 10 year period and divided by the rig capital costs. No inflation adjustments were performed. Average dayrates for jackup rigs over the period ranged from 0.05 to 0.19% of capital costs and appears to reasonably bound market data.[2] On average, dayrates constituted 0.11% of the capital costs of a rig.

Table 7.1 shows the costs of newbuild turbine installation vessels. Costs vary over time with the price of steel, shipyard competition, and vessel specifications, and range from $150 to $250 million for SPIVs. One liftboat at a cost of $50 million and two jackup barges each at $65 million have been built for the turbine installation market. The current trend is for SPIV construction and a few SPIVs approach or exceed $300 million.

Dayrates as a percentage of capital expenditures are shown in Table 7.2 for vessels ranging in capital cost from $30 to $300 million. Whether these costs are reflective of actual costs will depend on the similarity between the offshore wind market and the rig, liftboat, and OSV markets relative to market conditions, operating costs, rates of return, finance conditions, and other factors.

7.3 Vessel Leasing

In most offshore service markets, a third party builds, owns and operates a vessel, and leases it on a dayrate basis to a project developer. We use two methods to analyze the break-even costs of this strategy. First, we decompose vessel costs into finance costs, operating expenditures, and returns on investment and sum these components to estimate costs. We also build a net present value (NPV) model of the vessel investment decision.

7.3.1 Component Approach

Dayrates are determined by the supply and demand for vessel services, and in theory should reflect the cost of vessel operation. Costs are described in terms of

[2] For example, Superior Energy Services released information on newbuild and dayrate costs for its large (265 class) liftboats. Superior built two vessels in 2009 for a cost of approximately $25 million and leased these vessels for approximately 37,500 $/day. This gives a dayrate of approximately 0.15% of newbuild cost [5]. In its 2009 financial statement, Tidewater provided information on the newbuild cost of domestic and international deepwater platform supply vessels. For domestic vessels, the dayrate was 0.075% of newbuild costs while for international vessels the dayrate was 0.1% of newbuild costs [6].

Table 7.1 Capital costs of selected newbuild turbine installation vessels

Vessel name	Type	Year delivered	Shipyard	Buyer	Vessel cost (million $)	Source
Titan II	Liftboat	2009	Semco	KS Energy	50	[25]
NA	SPIV	2012	Aker (PA)	Bluewater	150	[12]
NA	SPIV	2012	Lamprell	Fred Olsen	160	[26]
MPI Discovery	SPIV	2011	Cosco	MPI	275	[27]
MPI Resolution	SPIV	2004	Shanaiguan	MPI	100	[28]
NA	SPIV	2011	Daewoo	RWE	140	[29]
Kraken and Leviathan	SPIV	2009	Lamprell	Sea Jacks	135	[30]
JB 114 and 115	Jackup	2009	Drydocks World		65	
Wind Lift 1	SPIV	2010	Western Shipyard	BARD	75	[31]
Sea Worker	Jackup	2008	Drydocks World	A2Sea	65	
Sea Installer	SPIV	2012	Cosco	A2Sea	140	[27]
Service Jack 1	SPIV	2010	Drydocks World	Master Marine	371	[32]
Seafox 5	SPIV	2012	Keppel FELS	Workfox	220	[27]

NA not available

Table 7.2 Turbine installation vessel dayrates as a percentage of capital costs

Capital cost (million $)	0.05%	0.10%	0.15%
30	15,000	30,000	45,000
50	25,000	50,000	75,000
100	50,000	100,000	150,000
150	75,000	150,000	225,000
200	100,000	200,000	300,000
300	150,000	300,000	450,000

three components: finance and capital costs, operating expenses, and returns on investment (ROI). Market conditions and levels of competition are time and region specific and are not considered even though their impacts may be significant and large. Taxes are assumed to be largely or entirely covered by vessel depreciation and are ignored. Estimates of finance costs, operating costs, and ROI for potential U.S. offshore wind installation vessels are made based on similar marine vessels; summing these components provides an estimate of the expected dayrate [7]. In Appendix A quantitative models of this development are presented.

7.3.1.1 Finance Costs

Vessels built in U.S. shipyards may be financed through private loans insured by the U.S. Maritime Administration [8]. These loans have a term up to 25 years at fixed interest rates 80–120 basis points above a Treasury bond with the same life to maturity[3] [9]. The choice of basis points will depend on the return required by the financier. Vessel construction often requires 20% of the cost of the vessel in cash or equity, however, in MARAD backed loans, the down payment can be as little as 12.5%. Vessel construction loans require a constant principal payment and declining interest payment over the life of the loan [8].

Assuming a $150 million vessel, 80% financed through a MARAD loan over 20 years at 5.5% would give an average amortized monthly payment of $776,145 ($25,871 per day). An additional cost of approximately $4,100 per day is added to recoup the non-financed 20% over the 20 year term. Given these assumptions, the total daily costs related to the recoup of capital expenditures would be approximately $30,000. This cost is then adjusted by dividing by the expected utilization rate[4] to determine the expected dayrate.

[3] For example, a vessel financed for 20 years and a yield on a 20 year security of 4.5%, gives an interest rate of 5.3% if 80 basis points are used and 5.7% if 120 basis points are used.

[4] Utilization rate is the proportion of time the vessel is under contract. For the future offshore wind industry in the U.S., utilization rate is unknown.

Table 7.3 Operating expenses for commercial vessels in the Gulf of Mexico (2009–2010)

Vessel type	Operating expense (1000 $/day)	Operating expense (% of dayrate)	Source
Jackup	32–40	42–58	[10]
Liftboat	5–13	62–64	[5, 33]
OSV	6–10	45–48	[6, 34]

7.3.1.2 Operating Expenses

Operating expenses denote all costs borne by a vessel operator beyond capital expenditures. The primary components of operating expense are labor, maintenance, fuel, insurance, and administrative costs. Operational expenses in dayrate contracts may be divided between the vessel owner and the client. For example, the vessel owner may pay crew expenses and maintenance costs while the client pays for fuel.

Typical operating expenses for liftboats, jackups, and offshore supply vessels (OSVs) in the GOM for 2009–2010 are shown in Table 7.3. Expenses are derived by dividing the annual costs of operation by the number of operating days. In general, operating expenses range between 40 and 65% of dayrates and vary from 5,000 $/day for small liftboats to 40,000 $/day for large drilling rigs. For wind installation vessels, operation expenses will likely lie within this range. Operating expenses are unlikely to fall below the operating expenses of liftboats because most TIVs are larger than liftboats and would require more maintenance and a larger crew. Operating expenses are also unlikely to exceed the operating costs of jackup rigs since drilling is one of the most complex and labor intensive activities in the offshore industry. The size and complexity of the installation vessel will determine its position within this range.

7.3.1.3 Returns on Investment

Jayaram and Royes [10] analyzed financial information of large international drilling rig operators from 1996 to 2008. They used the ratio of EBITDA[5] to replacement cost to derive an annual average ROI estimate. Over the study period, ROI ranged from approximately 4 to 14% and averaged 8.5%. Using data from the annual financial fillings of two large, publicly traded OSV operators (Tidewater and Hornbeck), we estimated returns in the GOM OSV market between 2005 through 2009. We used the ratio of net income to total assets to estimate the return on investment. Together, the two companies had an average ROI of 8.3%.

[5] EBITDA is earnings before interest, taxes, depreciation and amortization and is a commonly used measure of financial performance; it is particularly useful for firms with expensive long-lived assets such as ships.

Table 7.4 Break-even dayrates for leased vessels

			Dayrate ($/day)		Dayrate (% of CAPEX)	
Capital cost (million $)	Operating cost ($)	Utilization (%)	NPV method	Component method	NPV method	Component method
30	15,000	90	26,481	30,118	0.09	0.10
50	15,000	90	31,893	40,197	0.06	0.08
100	30,000	90	63,787	80,394	0.06	0.08
150	40,000	90	89,559	115,591	0.06	0.08
200	40,000	90	103,089	140,788	0.05	0.07
300	40,000	90	130,148	191,183	0.04	0.06
30	15,000	75	31,778	31,762	0.11	0.11
50	15,000	75	38,272	42,937	0.08	0.09
100	30,000	75	76,545	85,874	0.08	0.09
150	40,000	75	107,471	123,810	0.07	0.08
200	40,000	75	123,707	151,747	0.06	0.08
300	40,000	75	156,178	207,621	0.05	0.07

7.3.1.4 Dayrate Estimate

Total dayrate is the sum of the finance costs, operating expenses, and ROI. Table 7.4 shows dayrates under varying assumptions. In all cases the vessel is financed via a fixed principal payment loan. The loan covers 80% of the costs of the vessel and has an interest rate of 5.5%. While the overall dayrates range widely with vessel capital costs, the dayrates match relatively well with expectations. It is known that large liftboats, for example, which cost approximately $30 million to build, have had dayrates around $35,000, roughly similar to the dayrates estimated in the scenarios with roughly similar capital costs. Furthermore, vessels with $150 to $200 million capital costs are estimated to have dayrates between $115,000 and $150,000, roughly equivalent to recent jack-up rigs which have approximately similar capital costs and dayrates [11]. Estimated dayrates range from 0.06 to 0.11% of capital costs which is also roughly equivalent to known ratios.

A regression model of the input–output data is created by varying the input parameters and recording the output. The output is DAYRATE = DR ($/day) and the input variables include: CAPEX, the capital costs in million dollars; OPEX, the operating cost in dollars per day; TERM, the term of the loan in years; and UTILIZATION = U, the utilization rate in percent. The resulting model has the form:

$$\mathrm{DR} = 131,253 + 763 * \mathrm{CAPEX} + \mathrm{OPEX} - 2,125 * \mathrm{TERM} - 1,333 * \mathrm{U} \quad (7.1)$$

The regression is significant ($R^2 = 0.90$) and all parameters have the expected sign and are statistically significant. As CAPEX and OPEX increase, DAYRATE increases, and as the TERM and UTILIZATION increase, DAYRATE decreases. The fixed cost component of the model is large and positive. We might expect this component to be zero or nearly zero, suggesting that when all factors are set to zero, the dayrate is zero. Note, however, that when UTILIZATION is 100%, it is approximately equal to the fixed component, effectively canceling it out. Therefore, the intercept can be interpreted as being approximately zero with fixed costs accruing as the utilization rate decreases.

The regression coefficients are useful for assessing the sensitivity of dayrates to changes in parameters and to specify DAYRATE for any combination of input within the bounds of the model. Dayrates are particularly sensitive to changes in utilization rate and loan duration; e.g., a 1% increase in utilization rate decreases dayrates by $1,333, and a 1 year increase in the term decreases the dayrate by $2,125; a 1 million dollar increase in capital costs increases the dayrate by $763.

7.3.2 Net Present Value Approach

A net present value model was developed with assumptions similar to those used in the dayrate component estimate. A vessel operator is assumed to purchase and lease a vessel on a dayrate basis. The vessel is financed through a 20 year fixed principal-declining interest loan and the vessel depreciates at a constant rate over a 20 year period. Taxes are assumed to be 25% of net cash flow, minus depreciation expenses. The discount rate is 8%, and operating expenses, the cost of capital and utilization rates are identical to those used previously.

The NPV method gives similar but lower costs than the components approach as shown in Table 7.4. The primary difference between the two models is the components approach requires an annual return on investment while the NPV method discounts future cash flows. The NPV model includes taxes, however, taxes are not relevant because tax liability does not exceed the depreciation expense of the asset at break-even dayrates.

7.3.3 Limitations

These approaches assume that dayrates are set by the costs of owning and operating vessels. In reality, costs are set by supply, demand, pricing power, and other factors. In periods of low demand and excess supply, vessel owners would be expected to lease vessels at dayrates below vessel costs. In this case, the lower bound for dayrates is determined by the costs and benefits of "stacking" (deactivating and storing) the vessel; while stacking is a complex decision, vessel operators are unlikely to operate a vessel if the dayrate does not exceed the operating costs. Therefore, operating expenses set a lower bound on dayrates.

In periods of high demand, dayrates are limited by the developers' ability to pay. As demand for vessels exceeds supply, new vessels are ordered to meet demand, but there are delays between ordering and delivery of a vessel and during these periods dayrates increase. By themselves, an increase in vessel dayrates may have a modest impact on demand, but increasing vessel dayrates may be associated with a general increase in component costs, and as these costs rise, investment becomes unprofitable at marginal projects. As marginal projects are canceled or delayed, demand declines.

7.4 Newbuild Program

Instead of leasing a vessel, a developer may choose to newbuild and operate their own vessel. Two of the most advanced U.S. developers (Bluewater and Deepwater) are pursuing this route [12] and although this approach has not been common in Europe in the past, it has recently become more popular.[6] This approach has the advantage of reducing transaction costs [13], but in most cases the developer would not be able to employ the vessel over its entire lifetime. Therefore, a developer building their own vessel would need to either sell the vessel at the end of their project, or lease the vessel to other developers. Models for both alternatives are derived in Appendix B. We assume the developer sells the vessel at its depreciated value.

The total cost to a developer of using their own newbuilt vessel to install a wind farm is the total capital cost of the vessel, the finance costs paid between the time of purchase and sale, and the total operational costs, minus the depreciated sale price. Depreciation is assumed to occur at a constant 5% per year, but the actual sales price would be determined by market conditions [14]. The vessel dayrate is determined by dividing the total costs by the duration of the activity.

Table 7.5 shows the estimated dayrates. In each scenario, the interest rate was assumed to be 5.5% with 80% of the total vessel costs financed. Dayrates are composed largely of operating costs which comprise 36 to 69% of the total. Dayrates are lower than in the leasing option and range from 0.04 to 0.07% of capital expenditures.

A generalized regression model is developed for the DAYRATE expressed in $ per day and the input variables CAPEX, the capital costs in million dollars; OPEX, the operating cost in dollars per day; TERM, the term of the loan in years; and DURATION = DUR, the length of the project in years:

[6] Utility company DONG acquired the turbine installer A2SEA in 2009, BARD used its own vessel to construct the BARD 1 offshore wind farm, and RWE has recently acquired two new vessels.

Table 7.5 Estimated dayrates for a developer using their own newbuilt vessel, sold at the conclusion of the project

Capital expenditure (million $)	Operating cost ($/day)	Project length (years)	Dayrate ($/day)	Dayrate (% CAPEX)
30	15,000	3	22,455	0.07
50	15,000	3	27,425	0.05
100	30,000	3	54,849	0.05
150	40,000	3	77,274	0.05
200	40,000	3	89,699	0.04
300	40,000	3	114,548	0.04
30	15,000	6	22,184	0.07
50	15,000	6	26,973	0.05
100	30,000	6	53,945	0.05
150	40,000	6	75,918	0.05
200	40,000	6	87,890	0.04
300	40,000	6	111,836	0.04

Note Additional user input includes interest rate, term, percent financed and life of the vessel, assumed fixed at 5.5%, 20 years, 80%, and 20 years, respectively

$$\text{DR} = -1,016 + 242 * \text{CAPEX} + \text{OPEX} + 159 * \text{TERM} - 416 * \text{DUR} \quad (7.2)$$

The model describes the data with high correlation and all the coefficients are of the expected sign and statistically significant. DAYRATE is sensitive to operating costs and insensitive to most other parameters. For example, a one dollar increase in operating expenses increases dayrates by one dollar, but a one million dollar increase in capital costs increases dayrates by only $242. As the term of the loan increases, the dayrate increases. This occurs because as the term of the loan decreases, the monthly principal payment increases which causes the monthly interest payments. Therefore, the total interest paid on the loan decreases and the dayrate declines accordingly.

7.5 Dayrate Estimation in the U.S.

7.5.1 Assumptions

The three methods developed for estimating dayrates—as a proportion of capital costs, derived from leasing models, and for developer-owned vessels—are referred to as "proportion", "lease", and "build", respectively. We consider liftboats, jackup barges, and SPIVs as candidates for turbine installation vessels in U.S. waters.

We assume that liftboats will have a capital cost of approximately $25 to $50 million and operating expenses of 10,000–20,000 $/day; jackup barges will have

Table 7.6 Vessel dayrate range for three alternative models ($/day)

	Lease		Proportion		Build			
Vessel	Min	Max	Min	Max	Min	Max	Exp. value	Range
Liftboat	19,459	57,307	12,500	75,000	15,582	32,767	35,436	12,500–75,000
JU barge	33,919	99,614	25,000	150,000	26,164	50,534	64,205	25,000–150,000
SPIV	86,756	189,229	75,000	300,000	63,493	91,068	134,258	60,000–300,000

capital costs $50 to $100 million and operating expenses of 15,000–25,000 $/day; and SPIVs will have capital costs of $150 to $200 million and operating costs between 30,000–40,000 $/day.

We also assume that all loans are for 80% of the total vessel cost at 4.5 to 6% interest over 10–20 years. For the lease method, the expected return on investment is assumed to range from 7 to 9% and the expected utilization ranges from 75 to 100%. For the proportion method the vessel dayrate is assumed to range from 0.05 to 0.15% of capital costs. For the build method we assume that the vessel life is 20 years and the project duration is 2–5 years. We are conservative in our parameterization which increases the variance in our estimates but makes it more likely that vessel dayrates will fall within the specified range.

7.5.2 Results

Table 7.6 shows the results. The proportion method gives the largest range of values while the build method produces the least variable results. The overall expected dayrate range was in most cases set by the proportion method (i.e. both the largest and smallest values were derived from the proportion method). This suggests that if we were less conservative in our interpretation of the dayrate as a proportion of capital costs, the ranges would decrease significantly. The expected values for liftboats, jackups, and SPIVs are approximately $35,000, $64,000, and $134,000 per day, respectively.

The build method generally produced the lowest dayrates; this might indicate that the build method is the least expensive, however, the build method relies on a secondary market for the vessel; if all developers were to undertake the build method, there would be no secondary market and the build method would be much more expensive.

The average in Table 7.6 is based on the maximum and minimum values parameterized and does not weight alternative parameterizations by their likelihood. However, given what is known about vessel dayrates, the averages may be relatively close to the actual dayrates.

7.5.3 Limitations

The build method demonstrates lower costs than the leasing method but requires developers to undertake activities outside of their core business and introduces new risk. If one of the two methods becomes the dominant means of procuring vessels, the use of the other method is unwarranted and would bias the analysis. Since we do not know which method will prevail, both are considered.

The proportional method yielded the largest range of values, making it potentially less useful than the build and leasing models. However, it is the only empirically derived cost estimate and requires few assumptions. By contrast, both the leasing and build methods depend on a large number of assumptions. In many cases, the impacts of these assumptions on costs were examined through alternative parameterizations, but the models also make implicit assumptions about the methods vessel owners use to determine dayrates and the validity and impact of these assumptions is unknown.

In all three methods, dayrates or operating expenses from one market (e.g. the GOM liftboat or jackup market) are used to estimate costs in another market (the U.S. offshore wind market). This is necessitated by the nature of the problem but can lead to error. Most notably, error could occur through differences in contract structure. Offshore contracts may be on a turnkey or dayrate basis. In a turnkey contract, the vessel operator assumes engineering costs and risk and dayrates are consequently higher. In dayrate contracts, all logistic and engineering risk is held by the project developer [15]. Future U.S. wind contract structure has not yet been set, but the transfer of risk from the developer to the vessel operator will be associated with cost increases.

The dayrates estimated in this analysis are likely lower than the current dayrates in Europe. The Crown Estate [2] estimated dayrates for foundation and turbine installation vessels to be £130,000 to £150,000 (approximately $210,000 to $250,000 per day) for SPIVs with capital costs of £130 to £150 million (0.01% of capital costs). These costs are within the bounds of our estimates but exceed the average costs. This likely reflects the high demand for vessels in the European market. It is possible that the U.S. market may experience similar demand, or that supply constraints might increase U.S. prices, but in the near term we expect the U.S. market to experience lower prices than the European market due to a lower willingness to pay and moderately increasing demand.

7.6 Mobilization Costs

If installation vessels are moved between operating regions, mobilization costs must be considered. There is an assumption in the literature that mobilization costs for offshore wind are "high" [1, 16–19], however, there has been little quantification of these costs. Transport costs could be ignored if they are viewed by the contractor as permanent relocation of an asset to a new market. In the drilling rig market, long distance movements are often considered investment decisions rather than mobilization billed to a single project, and may be paid by the drilling contractor, operator, or both [4]. Mobilization cost is a function of the distance, vessel dayrate, vessel size, and transport method. Three methods for transport exist: self-propel, heavy-lift transport, or tow. Here, we derive formulas for the costs of mobilization.

7.6.1 Tug Transport

Tug transport would likely be used for derrick barges, heavy-lift vessels, and jackup barges. The cost of using a tug to pull an installation vessel is given by the time required (in days) multiplied by the cost per day plus the costs of insurance. Cost per day is composed of the tug dayrate, daily fuel use, and the installation vessel dayrate during transport. Formally:

$$\left[\left(\frac{X}{24V_t}\right) * \left(\left(0.011DV_t^2 * T\right) + \left(0.011DV_t^2 * G\right) + I\right)\right] + R_t \tag{7.3}$$

where X is the tow distance in nautical miles, V_t is the speed (knots) of the tug, D is the installation vessel displacement (tons), T is the tug dayrate per horsepower (\$/hp·day), G is the cost of fuel per gallon (\$/gal), I is the installation vessel dayrate (\$/day), R is the replacement cost of the installation vessel (\$) and N_t is the insurance premium (% of replacement cost) associated with wet tows. Both the tug dayrate and fuel use are a function of the tug horsepower (given by $0.011DV_t^2$; [20]) and the tug is assumed to use one gallon of fuel per horsepower per day.[7]

Displacement will typically range from 4,000 to 20,000 tons depending on the vessel specification. Tug dayrates per horsepower vary between 1.35 and 2.Vessel dayrate also depends on vessel specifications and if a lower dayrate for transport is offered. The insurance premium for a wet tow is typically 0.15–7.5% of replacement cost [21].

7.6.2 Self-Propelled Transport

Many turbine installation vessels are self-propelled and are capable of mobilizing without vessel support. In this case, the costs would be the travel time multiplied by dayrate plus fuel costs, or

[7] For dimensional analysis, set $0.011DV_t^2$ equal to tug hp and since fuel use is 1 gal per hp per day, G can be considered as \$/hp·day.

$$\left(\frac{X}{24V_i}\right)(I + (1.2 * P_i)G) \tag{7.4}$$

where V_i is the speed (knots) of the installation vessel and P_i is the installed power (hp). The variables *X*, *I* and *G* are the same as defined in Eq. (7.3). We assume that at cruise speed, 80% of the installed power is utilized [22]. The installed power typically varies from 4,000 to 14,000 hp and the speed ranges from 4 to 12 knots.

7.6.3 Heavy-Lift Vessel Transport

The cost of heavy transport is a function of the dayrates of the lift vessel and the installation vessel, the distance, fuel use, and insurance premium. The total cost is the number of days required for transport, times the daily cost, plus the insurance premium plus the dayrate during loading and unloading:

$$\left[\left(\frac{X}{24V_s}\right) * [S + I(1.2 * P_s)G]\right] + R * N_s + 2(S + I) \tag{7.5}$$

where V_s is the semisubmersible speed (knots), *S* is the semisubmersible dayrate ($/day), P_s is the installed power (hp) on the semisubmersible heavy-lift, and N_s is the insurance premium (%). *X*, *I*, *G* and *R* are defined in Eq. (7.3). The final term in the equation $(2(S + I))$ represents the additional cost required to load and unload a vessel on a semisubmersible vessel and assumes that one day is needed for unloading and loading.

Heavy-lift transport via semisubmersible[8] may allow for lower insurance premiums than a wet tow. Heavy-lift ships are also capable of transporting larger turbine installation vessels and derrick barges. Heavy-lift transport would likely begin to be favored over wet tows for distances over 1,000 to 2,000 miles.

For application, the dayrate for a semisubmersible heavy-lift vessel ranges from approximately $50,000 to $100,000 per day [23]. The installed power of semisubmersible lift ships (P_s) ranges from 16,000 to 27,000 hp and the insurance premium for semisubmersible transport is approximately 0.2% [21].

7.6.4 Example

The approximate mobilization distances between various locations along the U.S. Gulf and Atlantic coast is shown in Fig. 7.1. Each location is near an area that could become important for offshore wind energy. Based on these distances, we parameterize the models with five distances: 250, 500, 1000, 1500, and 2000 nautical miles.

[8] Heavy lift transport was used for the transfer of the *Titan I* and *Titan II* from the construction shipyard in Louisiana to the UK.

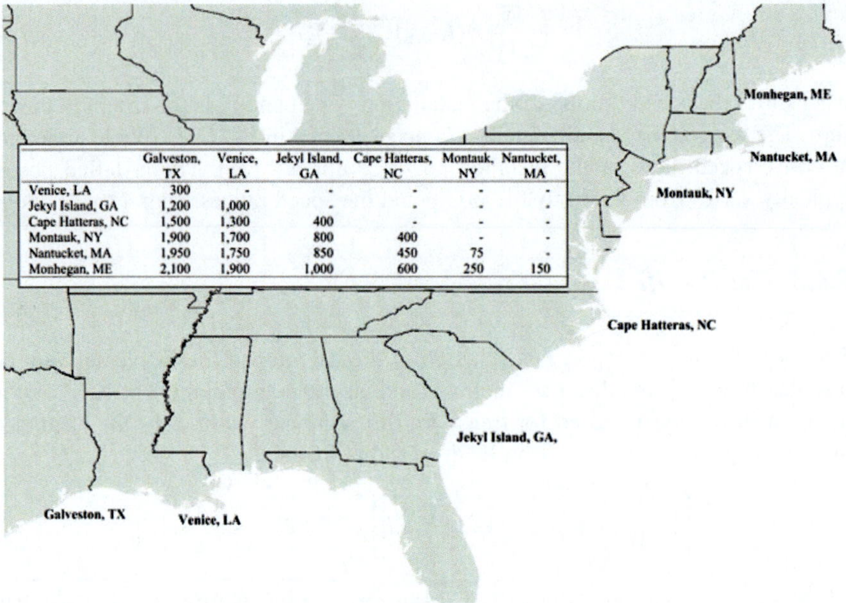

	Galveston, TX	Venice, LA	Jekyl Island, GA	Cape Hatteras, NC	Montauk, NY	Nantucket, MA
Venice, LA	300	-	-	-	-	-
Jekyl Island, GA	1,200	1,000	-	-	-	-
Cape Hatteras, NC	1,500	1,300	400	-	-	-
Montauk, NY	1,900	1,700	800	400	-	-
Nantucket, MA	1,950	1,750	850	450	75	-
Monhegan, ME	2,100	1,900	1,000	600	250	150

Fig. 7.1 Hypothetical mobilization locations and distances in nautical miles

We allow liftboats and large turbine installation vessels to be self-propelled but do not allow jackup barges to be self-propelled. We assume all fuel costs 3 $/gal; that liftboats and self-propelled vessels travel at four and ten knots, respectively; that vessels travel at seven knots while under tow; that liftboat displacement ranges from 4,000 to 10,000 tons; that barge displacement ranges from 10,000 to 15,000 tons; that self-propelled installation vessel displacement varies from 15,000 to 20,000 tons; that insurance premiums range from 0.15 to 3% for wet tows[9] and are 0.2% for dry transport; that installation vessel dayrates are equal to the averages given in Table 7.6; and that the replacement costs of vessels are equivalent to the newbuild costs. In cases in which the minimum costs of one mobilization method exceeded the maximum costs of another method, we assumed that the more expensive method would not be used and it was removed from the table. As a result, we do not consider liftboats to be transported by heavy-lift for 250, 500, 1000, or 1500 nautical mile distances, and we do not consider SPIVs to be transported by wet tow or heavy-lift. Results are shown in Table 7.7.

The ranges in Table 7.7 were determined as the minimum cost associated with the least expensive method, to the maximum costs associated with the next best method. The expected values are the averages of all values included in the range.

[9] Depending on the circumstances of the tow, wet tows may have insurance rates above 3%, however, this produces exceptionally high costs making wet tows extremely costly. If insurance premiums exceed 3% there is likely to be no case in which they are favored and we therefore exclude this possibility from the analysis.

Table 7.7 Ranges of mobilization costs by method of mobilization (1000$)

Vessel type	Distance (nm)	Tow Min	Tow Max	Self-propel Min	Self-propel Max	Heavy-lift Min	Heavy-lift Max	Range	Expected value
Liftboat	250	104	1,593	106	177	–	–	104–177	129
	500	171	1,686	213	353	–	–	171–353	246
	1000	304	1,871	425	707	–	–	304–707	479
	1500	437	2,057	638	1,060	–	–	437–1,060	712
	2000	570	2,243	851	1,413	1,413	2,310	570–1,413	1,061
JU barge	250	205	3,156	–	–	507	801	205–801	504
	500	335	3,311	–	–	686	1,073	335–1,073	698
	1000	595	3,623	–	–	1,044	1,618	595–1,618	1,086
	1500	855	3,934	–	–	1,402	2,162	855–2,162	1,473
	2000	1,115	4,246	–	–	1,760	2,707	1115–2,707	1,861
SPIV	250	–	–	151	234	–	–	151–234	192
	500	–	–	302	467	–	–	302–467	385
	1000	–	–	604	934	–	–	604–934	769
	1500	–	–	907	1,402	–	–	907–1,402	1,154
	2000	–	–	1,209	1,869	–	–	1,209–1,869	1,539

For example, in the first row of Table 7.7, the range is 104,000 to $177,000 but does not include $1,593,000 because we assume that if this cost is accurate it would never be selected due to the availability of cheaper options. In this case, three values are included in the range ($104,000, $106,000, and $177,000) and the expected value is the average of these three values. Jackups and liftboats have multiple options for mobilization while SPIVs have only a single alternative. This caused the ranges for jackup and liftboat mobilization costs to be greater than those for SPIV costs. In reality the existence of multiple transport options for liftboats and jackups might allow for competition and lower costs.

The costs associated with wet tows vary significantly due to unpredictable insurance costs [24]. Insurance rates are determined on a case-by-case basis and depend on a large variety of mostly nongeneralizable factors. If insurance costs are low, wet tows are generally the cheapest option for liftboats and jackup barges. However, when insurance costs are high, wet tows are never the most economic choice. Overall, mobilization costs range from approximately $100,000 to $3,000,000 depending on transport type, distance and vessel type. Relative to the total installation costs of a project, vessel mobilization is likely to be a small fraction in most cases.

7.7 Total Vessel Costs

Total vessel costs for a given project are the dayrate times the expected duration of the project, plus the mobilization cost. For illustration, assume a liftboat is built in Venice, Louisiana and is mobilized to Delaware for a project with an expected

Table 7.8 Total vessel cost for a one year installation project (1000$)

		With mobilization			Without mobilization		
Vessel type	Mob distance (nm)	Average	Min	Max	Average	Min	Max
Liftboat	250	13,063	4,667	27,552	12,934	4,563	27,375
	500	13,180	4,734	27,728	12,934	4,563	27,375
	1,000	13,413	4,867	28,082	12,934	4,563	27,375
	1,500	13,646	5,000	28,435	12,934	4,563	27,375
	2,00	14,245	5,133	28,788	12,934	4,563	27,375
JU barge	250	23,939	9,330	55,551	23,435	9,125	54,750
	500	24,133	9,460	55,823	23,435	9,125	54,750
	1,000	24,521	9,720	56,368	23,435	9,125	54,750
	1,500	24,908	9,980	56,912	23,435	9,125	54,750
	2,000	25,296	10,240	57,457	23,435	9,125	54,750
SPIV	250	49,197	22,051	109,734	49,004	21,900	109,500
	500	49,389	22,202	109,967	49,004	21,900	109,500
	1,000	49,774	22,504	110,434	49,004	21,900	109,500
	1,500	50,158	22,807	110,902	49,004	21,900	109,500
	2,000	50,543	23,109	111,369	49,004	21,900	109,500

duration of one year. From Table 7.6, the dayrate ranges between 12,500 and 75,000 $/day with an expected value of 35,436 $/day. From Fig. 7.1, the mobilization distance is approximated based on the starting location near Venice, LA and ending between Cape Hatteras, NC and Montauk, NY, approximately 1,500 nautical miles. From Table 7.7, the mobilization costs for a liftboat mobilized 1,500 nm ranges from $437,000 to $1,060,000 with an average value of $712,000. To find the total costs of the project we multiply the dayrate by expected duration of the activity and add the mobilization costs. In this case, the total vessel costs range from $5 to $28 million with an expected value of $13 million.

Table 7.8 summarizes the total costs of all vessel types and mobilization distances. To apply the table, select a vessel of interest and a desired mobilization distance. The table provides the total cost for one year of activity with and without mobilization costs. Demobilization costs are not included. The mobilization costs are typically 2–5% of the total costs for a one year project. While mobilization costs are not negligible, in most cases they are small compared to the total vessel costs and nearly negligible compared to the total costs for a large project.

Costs increase as the size of the vessel increases but these estimates do not include vessel spreads. Different vessel types will require different support spreads and larger vessels such as SPIVs will require smaller spreads, thereby compensating for their higher costs. Additionally, larger vessels may be able to complete a project in less time than a smaller vessel. The most efficient vessel choice will depend on the interaction of vessel cost, spread cost and size, and installation rates.

Appendix A. Leasing Dayrates

Dayrates associated with a vessel lease (D_L) are composed of three factors: the daily finance costs (F_d), the daily operating costs (O) and the daily ROI (R_d):

$$D_L = F_d + O + R_d \tag{7.6}$$

The daily finance costs are composed of the daily principal and interest payment to the lender plus the daily proportion of the down payment. Over the course of a term loan, the monthly interest payment declines with the declining principal; therefore the finance costs could be determined by either the average interest payment or the specific interest payment. For our purposes, the average interest payment is sufficient and simplifies the equations. In this case, the average monthly principal and interest payment (A) is the first payment plus the last payment divided by two:

$$A = \frac{\left[\frac{SP}{12T} + \frac{I}{12}SP\right] + \left[\frac{SP}{12T} + \frac{I}{12}\left(\frac{SP}{12T}\right)\right]}{2} \tag{7.7}$$

where S is the total ship cost, P is the proportion financed, T is the term of the loan and I is the interest rate. In Eq. (7.7), the two terms in square brackets in the numerator are the first and last payments, respectively. Within each bracketed term, the first term ($SP/12T$) is the monthly principal payment, while the second term is the monthly interest payment. The non-financed portion (N) of the capital costs are:

$$N = S - SP \tag{7.8}$$

The daily capital costs (F_d) are then:

$$F_d = \frac{A}{30.4} + \frac{N}{Y365} \tag{7.9}$$

where Y equals the life of the vessel in years (alternatively, T could be used instead of Y). To account for unutilized days, the daily finance costs can be divided by the expected utilization rate, U_e.

The operating costs, O, are derived from annual operating costs divided by utilized days and therefore already account for unutilized periods.

The daily ROI (R_d) is the total investment (S) times the annual ROI (R_A) divided by 365:

$$R_d = \frac{SR_A}{365} \tag{7.10}$$

R_d may also be divided by the expected utilization rate U_e.

The total dayrate is thus given by:

$$D_L = \frac{\left[\frac{SP}{12T} + \frac{I}{12}SP\right] + \left[\frac{SP}{12T} + \frac{I}{12}\left(\frac{SP}{12T}\right)\right]}{60.8 * U_e} + \frac{N}{Y365U_e} + O + \frac{SR_A}{365U_e} \tag{7.11}$$

Appendix B. Newbuilding Dayrate

The dayrate for newbuilding is derived differently from the dayrate for leasing. For newbuilding, the daily cost (D_N) to the developer is the total financed cost (F_T) divided by the duration of the project (L) in years plus the daily operating expense.

$$D_N = \frac{F_T}{L365} + O \tag{7.12}$$

This gives the vessel no value after completion of the project; to give value to the vessel we take the total cost of the project minus the total net income (G) from the vessel after the project, and divide by the duration of the project:

$$D_N = \frac{F_T + O365L - G}{L365} \tag{7.13}$$

In this case, G may be either a lump sum value accrued from the sale of the vessel (G_S), or it may be revenue generated from the future lease of the vessel (G_L).

F_T is given by the total vessel cost plus the interest paid where the total interest paid is calculated as the average of the first and last interest payments times the total number of interest payments:

$$F_T = S + T12 * \left[\frac{\frac{SPI}{12} + \left(\frac{SP}{T12}\right)\frac{I}{12}}{2}\right] \tag{7.14}$$

which simplifies to:

$$F_T = S + \frac{SPI}{2}\left(T + \frac{1}{12}\right) \tag{7.15}$$

Assuming the vessel is leased, the total vessel net income following use in the initial wind farm (G_L) is a function of the dayrate (not including operating costs), and the remaining life of the vessel:

$$G_L = 365(Y - L)U_e(D_L - O) \tag{7.16}$$

where Y equals the total expected life of the vessel, in years. The dayrate, D_L, is given in Eq. (7.11) and represents the net income for a similar vessel. Combining the equations and simplifying gives a total dayrate for a newbuilt vessel to be leased after initial use (D_{NL}) of:

$$D_{NL} = \frac{1}{L365}\left(\begin{array}{r} S + \frac{SPI}{2}\left(T + \frac{1}{12}\right) + O365L - \\ \left[365(Y - L) * \left(\frac{\left(\frac{SP}{12T} + \frac{I}{12}SP\right) + \left(\frac{SP}{12T} + \frac{I}{12}\left(\frac{SP}{12T}\right)\right)}{60.8} + \frac{N}{Y365} + \frac{SR_A}{365}\right)\right] \end{array}\right) \tag{7.17}$$

In Eq. (7.17), D_{NL} is negative for many plausible combinations of parameters which suggests that future earnings offset the costs of installation. However, a negative dayrate is uninformative for the purpose of cost estimation. Further, Eq. (7.17) assumes that the expected utilization rate equals the actual utilization rate and that vessel supply and demand are in equilibrium over the life of the vessel so that dayrates are well modeled by D_L. If supply and demand are not in equilibrium, D_L will be a poor estimator of the real dayrate.

Conversely, the vessel may be sold at the conclusion of the project, reducing future risk to the developer. The total revenue from the sale (G_S) is a function of the depreciation rate and the length of the project. Assuming that the vessel depreciates constantly over the vessel life (Y):

$$G_S = S * \left(\frac{Y - L}{Y}\right) \tag{7.18}$$

This gives a dayrate for a newbuilt vessel sold after completion of the project of:

$$D_{NS} = \frac{S + \frac{SPI}{2}\left(T + \frac{1}{12}\right) + O356L - S\left(\frac{Y-L}{Y}\right)}{365L} \tag{7.19}$$

However, it may be more reasonable to assume that when the vessel is sold the loan would be paid off, with no additional interest accruing. In this case, F_T is not a function of the total interest paid over the life of the loan, but the interest paid up until year L. In this case, we can determine the average interest payment as the first interest payment plus the interest payment at time L, divided by two. We then multiply the average interest payment by the number of interest payments made.

$$F_T = S + L \cdot 12\left(\frac{\frac{SPI}{12} + \left[SP - L12\left(\frac{SP}{12T}\right)\right] \cdot \frac{I}{12}}{2}\right) \tag{7.20}$$

This gives a total D_{NS} of:

$$D_{NS} = \frac{S + L \cdot 12\left(\frac{\frac{SPI}{12} + \left[SP - L12\left(\frac{SP}{12T}\right)\right] \cdot \frac{I}{12}}{2}\right) + O356L - S\left(\frac{Y-L}{Y}\right)}{365L} \tag{7.21}$$

We consider this to be the most useful equation for the costs associated with a newbuilt vessel.

References

1. Morgan CA, Snodin HM, Scott NC (2003) Offshore wind: economics of scale, engineering resource and load factors. Department of Trade and Industry, London. 3914/BR/01
2. Crown Estate (2010) A guide to an offshore wind farm. http://www.thecrownestate.co.uk/guide_to_offshore_windfarm.pdf
3. Cole JC, Herrmann RP, Scott RJ (1997) Discoverer Enterprise: World's most advanced drilling unit. In: SPE/IADC Drilling Conference, 4–6 March, Amsterdam, Netherlands. 37659-MS
4. Riglogix (2011). Database. www.riglogix.com, Accessed 15 Apr 2011
5. Superior Energy Services (2010) 2009 Quarterly report. http://ir.superiorenergy.com/phoenix.zhtml?c=97570&p=irol-sec
6. Tidewater (2010) 2009 Annual report. http://phx.corporate-ir.net/phoenix.zhtml?c=81406&p=irol-reportsannual
7. Cole JC (1995) Replacement cost economics of offshore drilling rigs. In: SPE/IADC Drilling Conference, February 28-March 2, Amsterdam, Netherlands. SPE 28197-MS
8. MARAD (2010) Title XI Federal Ship Financing Program. http://www.marad.dot.gov/ships_shipping_landing_page/title_xi_home/title_xi_home.htm
9. Esber B (2004) Title XI financing for vessel and shipyard improvement projects. MarineLink. http://www.marinelink.com/news/article/title-xi-financing-for-vessel-and-shipyard-improvement-projects/322695.aspx. Accessed 24 Oct 2010
10. Jayaram A, Royes A (2009) Offshore drillers: normalize it. Credit Suisse
11. Bailey JE, Sullivan CW, Cruise D (2010) Offshore drilling monthly. Jefferies and Company, Houston
12. Strauss-Wieder Inc (2010) Economic impact assessment. http://www.tigerii-tiv.com/. Accessed 20 Oct 2010
13. Jablonowski CJ, Kleit AN (2011) Transaction costs and organizational choice: modeling governance in offshore drilling. Eng Econ 56:28–58
14. Goulielmos AM (2009) Risk analysis of the Aframax freight market and of its new building and second hand prices, 1976–2008 and 1984–2008. Int J Shipp Transp Logist 1:74–97
15. Osmundsen P, Sorenes T, Toft A (2008) Drilling contracts and incentives. Energy Policy 36(8):3138–3144
16. Kempton W, Mandelstam P, Mercurio M, Musial W, Watson G, Uliman J, Stewart S (2009) Research and development needs for offshore wind. American Wind Energy Association, Offshore Wind Working Group
17. Thresher R, Robinson M, Veers P (2008) Wind energy technology: current status and R&D future. Presented at the Physics of Sustainable Energy Conference University of California, Berkeley. March 1-2. NREL/CP-500-43374
18. Green J, Bowen A, Fingersh LJ, Wan Y (2007) Electrical collection and transmission systems for offshore wind power. In: Offshore Technology Conference, April 30–May 3, Houston, 19090-MS
19. Musial W, Butterfield S, Ram B (2006) Energy from offshore wind: preprint. In: Offshore Technology Conference, Houston, NREL/CP-500-39450
20. Zahalka P (2008) Bollard pull. http://www.vht-online.de/vht2008/pdf/Bollard-Pull.pdf. Accessed 24 Oct 2010
21. Van Hoorn F (1990) Semi-submersible heavy-lift ships in operation. Schip en Werf 52(8):448–453
22. EPA (2000) Analysis of commercial marine vessels emissions and fuel consumption data. Washington, EPA420-R-00-002
23. De Jong E (2010) Ready for a heavy-lift of the share price. SNS Securities
24. Roddick BM (1995) Towage insurance. Tulane Law Review (70):679
25. Lim R (2008) KS Energy liftboat sinks on way to fulfill North Sea contract. The Business Times, 29 October

26. Gulfnews.com (2010) Lamprell Energy signs wind vessel deal. February 16. http://gulfnews.com/business/general/lamprell-energy-signs-wind-vessel-deal-1.583730. Accessed 20 Oct 2010
27. Philips P (2010) WTIV market expanding rapidly. Pacific Maritime online. http://www.pmmonlinenews.com/2010/10/wtiv-market-expanding-rapidly.html. Accessed 20 Oct 2010
28. Osborne A (2004) Mayflower installation ship sold for just £12 million. The London Telegraph, 22 April
29. RWE (2009) RWE awards contract to Korean shipyard to build an offshore construction ship. http://www.rwe.com/web/cms/en/86182/rwe-innogy/news-press/press/?pmid=4004290. Accessed 20 Oct 2010
30. Forston D (2009) Turbine ship proves to be an idea with legs. The Sunday Times. October 4
31. BARD Engineering (2008) Keel laying of Wind Lift 1. BARD Magazine. 6 June
32. Offshore Shipping (2008) Master Marine awarded first contract for self-propelled jackup. October 24. http://www.oilpubs.com/oso/article.asp?v1=7921. Accessed 20 Oct 2010
33. Hercules Offshore (2010) 2009 Annual report. http://phx.corporate-ir.net/phoenix.zhtml?c=192573&p=irol-reportsannual. Accessed 29 Oct 2010
34. Hornbeck Offshore (2010) 2009 Annual report. http://ir.hornbeck.com/phoenix.zhtml?c=132245&p=irol-reportsannual. Accessed 24 Oct 2010

Chapter 8
Capital Cost Estimation: A Reference Class Approach

8.1 Comparative Versus Reference Approach

Offshore wind capital cost estimates can be made through either an engineering (bottom-up) or comparative (top-down) approach. In an engineering approach, a model is developed that includes cost estimates for each component of the system, which when summed, yields an estimate of the capital cost of the wind farm [1]. The engineering approach is useful in estimating the costs of a particular project, but requires site-specific information and is subject to optimism bias [2].

In the comparative approach, cost data from existing projects are used as a basis for analogy. If the physical infrastructure in two regions is similar, then the project characteristics may be similar even if other characteristics of development—installation strategies, marine vessel fleet, government regulation, etc.—are different. The implicit assumption of this approach is that the commonalities of offshore wind projects associated with the technology, infrastructure, capital intensity, complexity, and installation requirements outweigh the differences due to environmental, contractual and market conditions.

A reference class is a set of existing projects for which cost information is used to improve the accuracy of comparative cost estimation by limiting the sample to those projects that are similar to a proposed project. A reference class approach can be used to estimate the costs of a specific project but here we develop a reference class of European offshore wind projects to inform cost estimates and uncertainty bounds of U.S. offshore wind farms. We assume the first offshore wind farms built in the U.S. will use monopile foundations and be placed in shallow ($\approx$20 m) water within 10 miles of the shore.

M. J. Kaiser and B. F. Snyder, *Offshore Wind Energy Cost Modeling*,
Green Energy and Technology, DOI: 10.1007/978-1-4471-2488-7_8,

8.2 Source Data

8.2.1 Sample Set

Capital expenditures were collected from trade journals, company websites, and academic and government reports. These values were compared using a commercial database [3] and an industry report [4]. Only wind farms that are operational and generating power, under construction, or where all capital contracts are finalized are considered. A total of 53 offshore wind farms are generating power or under construction as of May 2010 [5]. In addition, there are at least two wind farms for which contracts have been finalized (Lincs and London Array) which gives a total sample of 55 wind farms.

8.2.2 Exclusion

Wind farms were excluded from the analysis if there was no reliable information about their costs, if they were built before 2000, or if they were built in Asia. Projects installed before 2000 were excluded because they are primarily of a demonstration character, used small turbines, were sited in benign waters, and are not representative of current or expected future projects. Projects installed in Asia were excluded because the Asian market is currently very small and likely to have a different cost structure than the European market.

From 56 wind farms, 21 were excluded, leaving 35 projects in the total sample. In most cases (13 of 21), wind farms were excluded due to missing data; in four cases, wind farms were excluded due to their age, and in two cases Asian wind farms were excluded. Additionally, the Hywind project was excluded because the reported costs were nearly an order-of-magnitude higher than the average cost, and Avedore Holmes was excluded as it is not truly offshore.

8.2.3 Reference Class

A reference class of 18 farms was created by excluding all wind farms built before 2005 and projects that did not employ monopile foundations. The capital expenditures of both the total sample and the reference class were analyzed.

8.2.4 Adjustment

Costs are adjusted to a single currency (the U.S. dollar) at a single time (1 January 2010) to allow meaningful comparisons. Due to widely varying exchange rates in

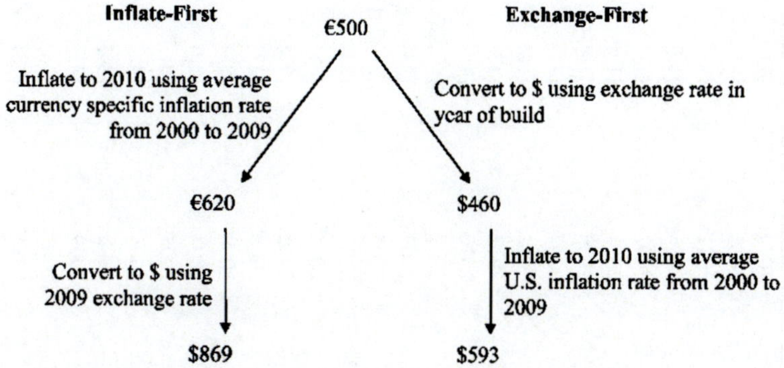

Fig. 8.1 Diagrammatic depiction of alternative methods of adjusting costs for a hypothetical €500 million wind farm built in 2000

the sample period, the order in which currency conversion and inflation adjustment are performed will impact comparisons. Two options can be employed.

Project costs could be inflated from the year of construction to the present in the reported currency (e.g., euro), then exchanged to U.S. dollars (called inflate-first). Alternatively, project costs could be exchanged to dollars using the exchange rate at the time of construction, and then inflated using the U.S. inflation index (called exchange-first). See Fig. 8.1. We use the inflate-first method and illustrate the options with an example in Appendix A. Inflation rates are based on a 10-year average. Exchange rates are based on the average exchange rate in the fourth quarter of 2009.

8.2.5 Normalization

All projects were normalized by nameplate capacity and price expressed in $/MW. Differences in the scope of project costs were taken into account where information was available. For example, the total project cost for Scroby Sands was reported as £80.1 million, but 8.5% of the budget included a five year O&M component; this portion was removed and the final adjusted cost was £73.2 million [6].

8.3 Capital Expenditures

8.3.1 Summary Statistics

Table 8.1 shows the nominal capital costs of all offshore wind farms in the sample. In Table 8.2, the capital costs are depicted along with estimates from a commercial database (4C Offshore) and an industry source (Garrad Hassan; GH). For our

Table 8.1 Capital costs of offshore wind farms in the total sample (2010)

Wind farm	Status	Capacity (MW)	Cost (million $)	Currency	Year online	Source
Alpha Ventus	Generating power	60	250	Euro	2009	[7]
Arklow	Generating power	25	45	Euro	2005	[8]
Bard I	Under construction	400	1400	Euro	2011	[9]
Barrow	Generating power	90	123	GBP	2006	[10]
Beatrice	Generating power	10	35	GBP	2007	[11]
Belwind1	Under construction	165	614	Euro	2011	[12, 13]
Blyth	Generating power	4	4	GBP	2000	[14]
Burbo Bank	Generating power	90	181	Euro	2007	[15]
Global Tech I	Under construction	400	1200	Euro	2012	[16]
Greater Gabbard	Under construction	504	1300	GBP	2012	[17]
Gunfleet Sands	Generating power	172	3900	DKK	2010	[18–21]
Horns Rev	Generating power	160	278	Euro	2002	[6, 22]
Horns Rev II	Generating power	209	3900	DKK	2009	[23]
Kentish Flats	Generating power	90	105	GBP	2005	[21, 24]
Lillgrund	Generating power	110	1800	SEK	2007	[21]
Lincs	Contracts signed	270	725	GBP	2012	[25]
London Array	Contracts signed	630	2200	Euro	2012	[26]
Lynn/Inner Dowsing	Generating power	194	300	GBP	2008	[27]
Middelgrunden	Generating power	40	44.9	Euro	2000	[28]
North Hoyle	Generating Power	60	82	GBP	2003	[29]
Nysted	Generating Power	165	250	Euro	2003	[6]
OWEZ	Generating power	108	217.7	Euro	2006	[30]
Princess Amalia	Generating power	120	383	Euro	2008	[24]
Rhyl Flats	Generating power	90	190	GBP	2009	[31]
Robin Rigg	Generating power	180	420	Euro	2008	[26]
Rodsand II	Under construction	207	400	Euro	2011	[26]
Samso	Generating power	23	35	Euro	2003	[26]

(continued)

Table 8.1 (continued)

Wind farm	Status	Capacity (MW)	Cost (million $)	Currency	Year online	Source
Scroby Sands	Generating power	60	73	GBP	2004	[6]
Sheringham Shoal	Under construction	317	10000	NOK	2011	[32]
Thanet	Under construction	300	780	GBP	2010	[22]
Thornton Bank I	Generating power	30	150	Euro	2009	[33]
Thornton Bank II&III	Under construction	295	1300	Euro	2013	[34]
Utgrunden	Generating power	10.5	14	Euro	2000	[35]
Walney	Contracts signed	367	8716	DKK	2011	[36, 37]
Yttre Stengrund	Generating power	10	13	Euro	2001	[38]

Table 8.2 Comparison of normalized capital costs by source (million $/MW)

Wind farm	Authors	4C [3]	GH [4]	Average
Alpha Ventus	6.1	6.1	6.6	6.2
Arklow*	2.9	2.9		2.9
Bard I	4.9			4.9
Barrow*	2.4	2.7	2.7	2.6
Beatrice	6.0	6.0		6.0
Belwind*	5.2	2.7		3.9
Blyth	2.0	2.0	2.2	2.1
Burbo Bank*	3.1	3.1		3.1
Global Tech I	4.1	4.5		4.3
Greater Gabbard*	3.9	4.8	6.0	4.9
Gunfleet Sands*	4.4	2.8	4.2	3.8
Horns Rev	2.9	2.9	2.1	2.6
Horns Rev II*	3.7	3.3	4.2	3.7
Kentish Flats*	2.1	2.4	2.3	2.3
Lillgrund	2.4	2.7	2.1	2.4
Lincs*	4.1	4.1		4.1
London Array*	4.8	4.8	4.8	4.8
Lynn/Inner Dowsing*	2.6	2.5	3.1	2.7
Middelgrunden	2.0	2.1	1.6	1.9
North Hoyle	2.6	2.5	2.4	2.5
Nysted	2.5	2.5	1.7	2.2
OWEZ*	3.1	2.9	2.7	2.9
Princess Amalia*	4.8	4.4	3.6	4.2
Rhyl Flats*	3.5	3.6	4.4	3.8
Robin Rigg*	3.5	3.6	3.7	3.6
Rodsand II	2.7	2.8		2.7
Samso	2.5	2.2	1.4	2.0
Scroby Sands	2.2	2.3	2.1	2.2
Sheringham Shoal*	5.2	5.2	4.9	5.1
Thanet*	4.1	4.5	5.6	4.8
Thornton Bank I	7.3	3.9	7.3	6.2
Thornton Bank II &III	6.2			6.2
Utgrunden	2.4	2.2	2.0	2.2
Walney*	4.5	5.1	4.6	4.7
Yttre Stengrund	2.3	2.3	1.7	2.1
Average (SD)—All	3.7 (1.4)	3.4 (1.2)	3.5 (1.7)	3.6 (1.4)
Average (SD)—Reference class	3.7 (1.0)	3.6 (1.0)	3.9 (1.2)	3.8 (0.9)

Note * Denotes element of reference class. Standard deviation depicted in parenthesis

estimates and the 4C Offshore data set, the inflate-first method was used. For the GH data, entries were normalized, inflated to 2009 prices, and adjusted to pounds by Garrad Hassan; we then converted to dollars using the 2009 exchange rate.

In most cases the values in the three datasets are similar or identical; however, in a few cases the values diverge significantly. Similarity does not imply reliability

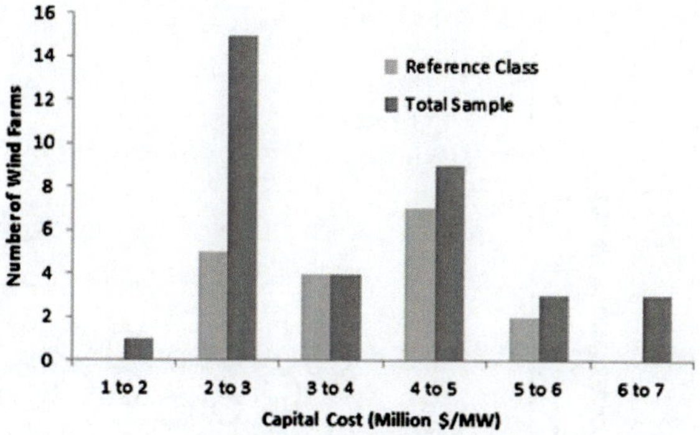

Fig. 8.2 Histogram of capital costs

as the source of the data is likely to be the same. Differences may be due to the time in which the source estimate was published (i.e. before or after all contracts are finalized), the degree of rounding, the scope of the source,[1] or methods of adjustment and inflation. We use the average of the three data sets for all subsequent analyses.

The histogram of the costs for the total sample and reference class is depicted in Fig. 8.2. The average cost of the total sample and reference class was $3.7 million/MW and $3.9 million/MW, respectively. For wind farms built after 2010, the average capital expenditures (CAPEX) is $4.7 million/MW in the total sample and $4.8 million/MW in the reference class.

The average capital expenditures of the reference class is slightly larger than the total sample, while the standard deviation of the reference class is lower than the total sample (1.4 vs. 0.9). When the reference class is restricted to projects online after 2010, the standard deviation declines further. A two standard deviation interval gives an expected range of capital costs from 2.1 to 5.7 million $/MW for all projects in the reference class and from 4 to 5.6 million $/MW for projects online after 2010.

8.3.2 Time Trends

Historical trends for adjusted, normalized offshore wind capital expenditures are shown in Fig. 8.3 and Table 8.3. Trends are shown only for the total sample. The average cost of offshore wind installation has increased from 2.2 million $/MW between 2000 and 2004 to over 4 million $/MW for recent developments. Figure 8.3 also shows that price increases have occurred as wind farms have increased in capacity. Many industry observers expected capacity increases and learning to lead

---

[1] For example, in the Thornton Bank project we used the cost of the first 30 MW phase, while the 4C data reported the estimated cost of the full 300 MW development.

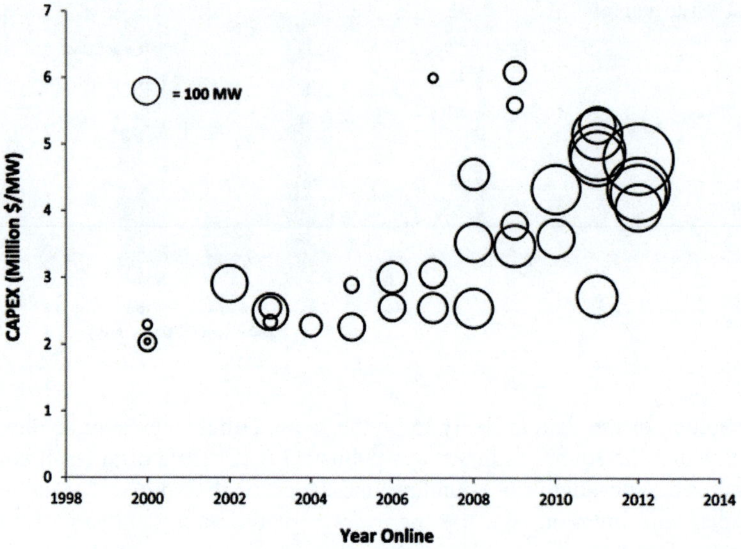

Fig. 8.3 Offshore wind farm capital expenditure and time of contract for total sample. Capacity expressed in bubble form

Table 8.3 Offshore wind farm capital expenditure by year of initial operation

Year online	CAPEX (million $/MW)	Number in dataset
2000–2004	2.2	9
2005–2007	3.2	7
2008–2010	4.4	9
2011+	4.7	10

to a reduction in offshore wind costs through economies of scale, but factors forcing costs upward in recent years have had a greater influence. Reasons for the increase in project costs have been attributed to several factors including increasing water depths, increasing commodity costs, reduction and centralization of supply chain competition, and increasing demand from onshore wind [39].

8.3.3 Economies of Scale

The impact of scale economies on development cost is difficult to assess since wind farms increased in size over the past decade as prices increased. Table 8.4 shows the cost of wind farms by generation category. There is little variation in wind farm costs by capacity, but wind farms over 250 MW are generally more expensive than smaller farms, suggesting that economies of scale do not currently

Table 8.4 Average capital expenditure by installed capacity

Project type	Capacity (MW)	CAPEX (million $/MW)	Number in dataset
Demonstration	<20	3.1	4
Pre-Commercial	20–100	3.3	11
Small commercial	100–250	3.3	11
Full commercial	250–750	4.9	9
Large commercial	>750		0

Table 8.5 Offshore wind farm capital expenditure by capacity and year online (million $/MW)

Year online	<20 (MW)	20–100 (MW)	100–250 (MW)	250–750 (MW)
2000–2004	2.1 (3)	2.2 (4)	2.4 (2)	
2005–2007	6.0 (1)	2.7 (4)	2.7 (2)	
2008–2010		5.5 (3)	3.6 (5)	4.8 (1)
2011+			4.0 (2)	4.9 (8)

Note Sample size denoted in parenthesis

govern development. In Table 8.5, the cost of offshore wind farms is presented by capacity and year online. Comparing costs within a time period controls the effects of time. Comparing across rows, there is no definitive trend of scale economies; however, sample sizes are too small for statistically meaningful conclusions.

8.3.4 Regression Model

Regression models of normalized capital costs were constructed to investigate the physical features that impact capital expenditures. Models were based on the linear form:

$$C = \alpha_0 + \alpha_1 \text{ CAP} + \alpha_2 \text{ WD} + \alpha_3 \text{ DIS} + \alpha_4 \text{ GRAV} + \alpha_5 \text{ JAC} + \alpha_6 \text{ STEEL}$$

where cost C is reported in million dollars per MW and explanatory variables included installed capacity (CAP, MW), water depth (WD, m), distance to shore (DIS, km), and a European steel price index lagged 2 years[2] (STEEL). Two indicator variables for gravity (GRAV) and jacket or tripod (JAC) foundations were also included. The number of turbines and year of installation were not considered because of multicollinearity. No interaction terms were evaluated because of the limited size of the sample and inherent constraints on the predictive ability of the variables.

[2] Steel prices were lagged because there is a significant delay between the time at which a contract is let and the time the project comes online. For example, 2006 steel prices would be used to estimate costs for a wind farm online in 2008.

Table 8.6 Summary of capital cost regression models

Cost = $\alpha_0 + \alpha_1$ CAP + α_2 WD + α_3 DIS + α_4 GRAV + α_5 JAC + α_6 STEEL

	Model	α_0	α_1	α_2	α_3	α_4	α_5	α_6	R^2
Normalized cost (million $/MW)	A	0.73	0.0011	0.036*	-0.0036	0.076	1.59*	0.013*	0.66
	B	0.63		0.038*	−0.0002	0.023	1.35*	0.014*	0.66
	C	0.75	0.0009	0.033*			1.46*	0.013*	0.68
	D	0.64		0.037*			1.35*	0.014*	0.67
	E	3.03*	0.0031*						0.16
	F	2.18*		0.067*					0.51
	G	3.03*			0.0241*				0.28
	H	0.67*						0.020*	0.34
Total cost (million $)	I	−110.01*	4.59*	5.66*	0.57	−94.75	27.35	−0.70	0.97
	J	−222.11*	4.55*	6.49*					0.97
	K	−121.57*	4.74*						0.96

Note * Statistically significant ($p < 0.05$)

Model results are given in Table 8.6. All of the models are statistically significant and have the expected signs for the coefficients. Models A through D explain similar proportions of the variance; however, at least one of the coefficients in Models A to C are not significant; therefore, Model D—which contains water depth, steel price and a jacket/tripod indicator variable—is the preferred model. The indicator variable for jackets and tripods is a better predictor of cost than the gravity foundation indicator variable. The gravity indicator coefficient was never significant; by contrast the indicator variables for jackets and tripods were usually significant. Table 8.6 also shows models for total capital costs. Total cost models explain more of the variance in costs but are poorly suited in evaluating the impacts of other site-specific variables.

Figures 8.4, 8.5, 8.6, and 8.7 show the results of the single variable regressions described in Models E to H. In each case, the models are significant, but do not generally predict a significant portion of the variance.

In Fig. 8.4, there is a slight positive relationship between capacity and capital expenditures; if economies of scale were present, this relationship would be negative.

In Fig. 8.5, there is a statistically significant relationship between water depth and capital costs; the relationship explains half of the variance in costs. While this is of limited utility as a basis for cost estimation, it does illustrate the importance of water depth on costs.

Figure 8.6 shows the influence on distance to shore. When two outlying data points associated with German wind farms (BARD I and Global Tech I) are removed, the model fit increases to 0.49

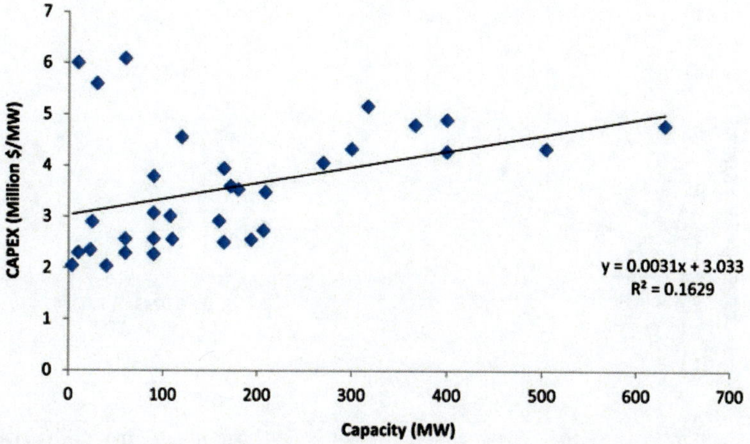

Fig. 8.4 Relationship between installed capacity and normalized capital expenditures

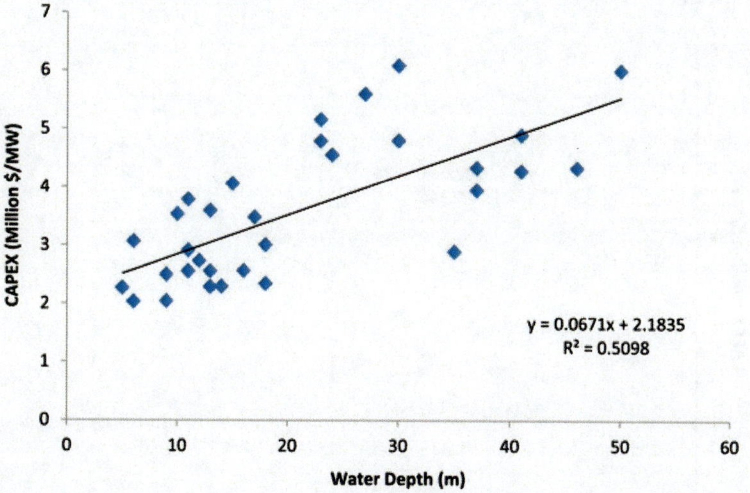

Fig. 8.5 Relationship between water depth and normalized capital expenditures

Figure 8.7 shows the influence of steel prices on costs. Steel price is a reasonable predictor of costs because of the significant role of steel in capital expenditures. Varying the time lag between steel price index and online date did not significantly modify the results.

In Fig. 8.8, the relationship between capacity and capital expenditures is depicted. Capacity is a good predictor, indicating that capital costs may be reasonably estimated by simply multiplying the average cost per MW by the project capacity.

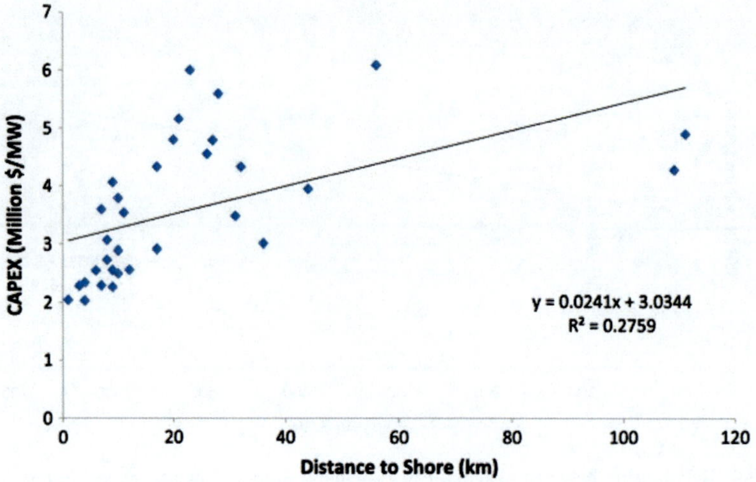

Fig. 8.6 Relationship between distance to shore and normalized capital expenditures

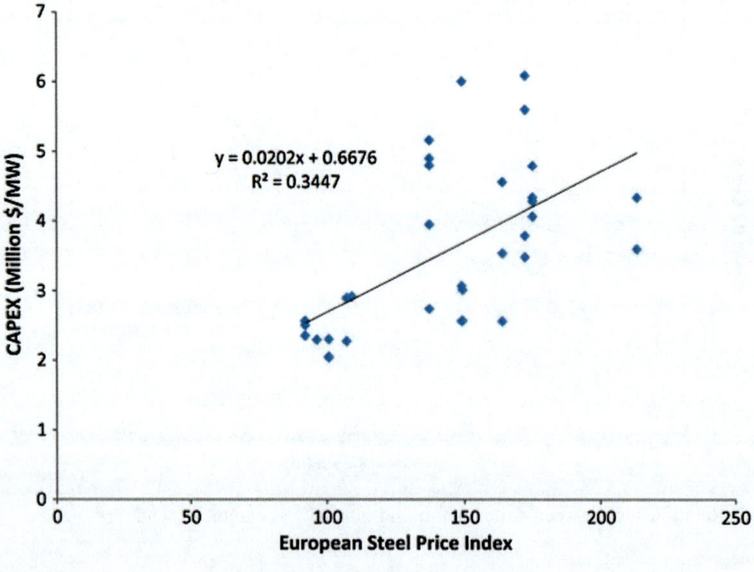

Fig. 8.7 Relationship between European steel price index and normalized capital expenditures

8.4 U.S.-European Comparisons

8.4.1 Turbines

The costs of offshore turbines are a primary driver of capital costs. Turbine costs are a function of supply and demand in regional markets, raw material costs, and

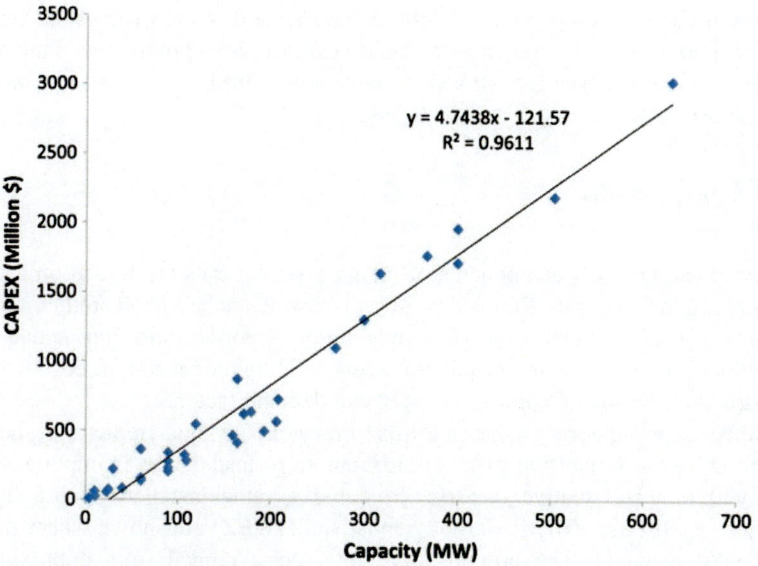

Fig. 8.8 Relationship between capacity and capital expenditures

transport cost. In 2010, the U.S. has a well-developed onshore turbine fabrication industry [40] but no capacity for offshore turbine manufacturing. U.S. developers plan to use turbines imported either from Europe or from China and are expected to have roughly similar costs.

8.4.2 Foundations

The costs of turbine foundations will be principally influenced by steel and labor costs. The capital and infrastructure requirements for monopiles are minimal and they are likely to be domestically sourced. The costs of foundations will also respond to the demand for foundation construction and the supply of construction services. Manufacturing wages in Western Europe are higher than in the U.S. [41], but differences vary with geography. During 2010, European steel prices were slightly lower than North American prices [42]. Foundation prices are therefore expected to be broadly similar in Europe and the U.S.

8.4.3 Cable

Cable costs will vary between the U.S. and Europe. The U.S. lacks the high voltage cable manufacturing facilities necessary for offshore wind farms and cable will likely be imported. However, these cables can be heavy and transport costs may be high. Depending on the weight, it is possible that some cables may need to be transported from Europe on specialized vessels rather than in break bulk; if this is the case, cable costs may be significantly higher in the U.S. than in Europe.

8.4.4 Installation

Many of the physical components of offshore wind farms are traded on a global market and in these cases European costs are expected to be a good predictor of U.S. costs. However, installation services may not be imported from Europe due to the restrictions of the Jones Act. Installation costs will be a function of the costs of vessel construction, personnel costs and supply and demand factors.

Since the installation market in Europe may not be a good predictor of the costs in the U.S., the proportion of costs attributable to installation is important since this will indicate relative impacts. In Table 8.7, the installation cost at three offshore wind farms (Blyth, Scroby Sands, and OWEZ) are shown where reliable data was available. The proportion of total costs ranged from approximately 10–30% and was highest at Blyth, an early small-scale wind farm.

The last six rows of Table 8.7 show component installation costs at several different wind farms. Total installation costs as a proportion of capital costs are the sum of turbine, foundation and cable installation. The data suggests that each of these activities represents approximately 3–6% of capital costs with cable installation being the least expensive and foundation installation being the most expensive. Several generic[3] estimates of offshore wind installation costs also exist (Table 8.8) with estimates ranging from 10 to 22% of capital costs.

Taken together, these data suggest that installation costs make up on the order of 20% of capital expenditures in European wind farms. As a result, for every 10% difference in installation costs in Europe and the U.S., total capital expenditures will change by approximately 2%; therefore, installation costs in the two regions can be different without major impacts on capital costs.

8.4.5 Site Selection

There are a number of factors that make offshore wind more attractive in Europe than the U.S. These include government involvement, higher average wind speeds, limited onshore renewable energy opportunities, high electricity prices, and public support of offshore wind. Because of these factors, European developers may be able to justify the development of sites that are more challenging and costly than in the U.S. and this may make the average U.S. costs lower than those in Europe.

[3] Generic estimates are based on model results or industry surveys rather than actual data.

Table 8.7 Component cost estimates of offshore wind farms (no inflation adjustment)

Wind farm	Scope of work	Unadjusted cost (million)	Year	Proportion of total cost (%)	Source
Blyth	Installation of piles and turbines	1.2 £	2001	31	[14]
Scroby Sands	Offshore installation	16.7 £	2004	23.4	[6]
OWEZ	Installation, including transport	42 €	2008	21	[30]
North Hoyle	Install 30 monopiles	5 £	2002	6	[43]
Thanet	Install infield and export cables	27 £	2008	3	[44]
Robin Rigg	Install export cable	7 £	2008	2	[4]
Greater Gabbard	Install turbines (14 month contract)	62 $	2009	3	[4]
Walney	Install turbine (18 month contract)	79 $	2009	5	[4]
Sheringham Shoal	Install 88 turbines and 2 substation modules	78 €	2009	6	[45]

Table 8.8 Offshore installation cost estimates

Source	Installation proportion (%)	Method
[46]	22	Generic model
[47]	19	Generic model
[5]	19	Industry experience
[48]	16	–
[49]	7*	Generic model
[50]	18	Generic model
[51]	9.6	Project budget

Note * Only includes turbine installation

8.5 Cost Drivers

8.5.1 Economic Recession

The global economic recession that began in 2008 had major impacts on the demand for onshore and offshore wind farm components. A major effect of the global financial crises was a tightening in credit markets and an increase in internal investment criteria. This not only impacts the ability of offshore wind developers to finance projects, but also affects onshore development. Since high onshore development constrains the offshore supply chain, the tightening of the credit market will lead to lower capital expenditures, and finance costs may increase.

8.5.2 Commodity Prices

Copper, steel and oil prices play a role in determining the costs of offshore wind development. Copper is primarily used in cables, transformers, and other electrical components of wind turbines, and is estimated to account for 3% of the total capital expenditures [5]. Therefore, while copper prices can be highly variable, even large changes in copper prices may have modest impacts on overall costs. Steel is estimated to make up 10–15% of the total costs of offshore wind projects, but steel prices have been less variable than copper prices. As a result, steel price variation is a more important cost driver than copper prices, but has not been a major driver of overall costs.

Energy prices also influence offshore development costs, however, their effects are largely indirect. Oil prices influence the installation costs of offshore wind power through competition for marine construction services. To the extent that offshore oil and offshore wind share the same construction supply chain, oil prices will continue to be an important cost driver. However, the demands of the two industries are different and as offshore wind develops specialized offshore wind firms may become dominant; in this case, the influence of oil prices on offshore wind capital costs would decline.

Coal and natural gas prices also have an impact on offshore wind development costs. Coal and natural gas are used in steelmaking, but more importantly, investors' expectations of future energy prices may influence the decision to build onshore and offshore wind farms. As coal and natural gas prices rise, wind energy becomes more attractive and demand grows. This tightens the supply chain and increases costs. Conversely, as coal and natural gas prices stabilize or decline, demand for wind turbines declines and prices fall.

8.5.3 Supply Chain

The supply chain for turbines, foundations, cables and installation services are all important cost drivers. Wind turbine supply accounts for the largest single cost component of offshore wind farms and even small changes in turbine costs have large impacts on capital expenditures. Offshore developers must compete with onshore developers for access to turbines and turbine components and turbine supply has been limited as onshore development expands. The offshore turbine supply market is highly concentrated with Siemens, Vestas, RePower, and Areva being the primary players; however, several other firms including GE, Nordex and Clipper Wind power are expected to enter the market in the coming years, potentially increasing supply and competition and lowering capital costs.

The supply of installation services is limited by the number of main installation vessels available. Vessel construction is a long-term investment decision and requires high utilization over a long period to warrant investment. Given the emerging state of the industry, it can be difficult to justify new building. Additionally, if investment in new vessels is justified, there may be long delays between orders and deliveries and these factors can result in periods in which supply is inadequate and prices rise.

8.6 Previous Estimates

Several recent studies have estimated the costs of near-term offshore wind farms in Europe. In 2009, Ernst and Young [52] estimated capital expenditures for European farms as £3.2 million/MW; BWEA and Garrad Hassan [5] estimated capital expenditures at £3.1 million/MW. These values are equivalent to \$4.3–\$5.4 million per MW in 2010 dollars, consistent with our estimates for wind farms online after 2010.

Several U.S. developers and consultants have released estimates of capital expenditures for U.S. offshore wind farms. Coastal Point Energy, the developers of the Galveston offshore wind farm, estimates its development will cost 3.5–4 million $/MW [53]. Weiss and Chang [54] estimated Cape Wind's capital costs at 5.6 million $/MW. Deepwater Wind has estimated costs for its Block Island wind farm at $250 million for a 30 MW project (7,300 $/MW). Our near-term estimate of 4.8 million $/MW falls between the Galveston and Cape Wind estimates.

8.7 Model Limitations

8.7.1 Sources of Error and Bias

Sample size. The database is limited in number and diverse in terms of project size, ownership, geographic region, year of construction, operating status, and foundation type. The diversity helps to ensure broad coverage of development, but the small sample precludes robust regression models.

Data reliability. Ideally, capital cost would be reported using uniform and consistent accounting categories across all projects, however, this does not occur in practice. Much of the data comes from press releases which are not specific about what is or is not included in capital costs. Press release data may or may not include grid interconnection costs, costs of capital, initial operating costs and state subsidies. In some cases, high quality reports with detailed cost accounting are available. In other cases, much less information is available. It is possible that the frequency and quality of reports is biased based on the size, novelty or developers of the wind farm, which could bias cost estimates. It is also possible that a data source report an estimated cost rather than an actual cost.[4] The overall impact of these issues may vary from minor to significant.

Contract type and currency. Contract type is an important determinant of project cost and the currency in which the contracts are reported may not be the currency in which the contracts are let. This ambiguity may lead to estimation variance.

Exchange rate fluctuations. Offshore wind projects have been performed over ten years in several countries and capital costs have been reported in several currencies. Exchange rates fluctuate over time and inflation rates are currency specific. To allow comparisons, all monetary values must be converted to a standardized format; the method of this conversion can cause errors or bias.

Normalization. Offshore wind projects are constructed in different environmental conditions and water depths, based upon different technologies and marine vessel spreads, and under different contract requirements. Collectively, these differences

---

[4] For example, in December 2009 the London Array Project was reported to have finalized contracts at a cost of €2 billion. However, in February 2010, the London Array Consortium signed additional contracts increasing the estimated price to €2.2 billion.

Table 8.9 Impact of alternative methods for adjusting costs

Year of completion	Exchange rate	Exchange-first (million 2010$)	Inflate-first (million 2010$)	Percent difference
2000	0.92	592.9	866.8	46.2
2001	0.89	559.2	848.4	51.7
2002	0.95	581.9	830.5	42.7
2003	1.13	674.8	813.0	20.5
2004	1.24	722.0	795.8	10.2
2005	1.24	703.9	778.9	10.7
2006	1.25	691.8	762.5	10.2
2007	1.37	739.2	746.3	1.0
2008	1.47	773.3	730.6	−5.5
2009	1.4	718.0	715.1	−0.4

create differences in cost. The primary normalization variable is generation capacity; comparisons using a multi-dimensional approach are preferred but limited by the size of the data set. Interaction effects need to be considered carefully.

Expensing costs. Project costs may be carried by the developer as overhead. For example, a developer may include support staff or management salaries, facility or equipment costs as overhead. This type of error is likely to produce bias in project comparisons.

8.7.2 Reference Class Constraints

The top–down approach to capital cost estimation has advantages and limitations. The primary rationale for using the technique depends upon the following arguments:

- The infrastructure, technologies, and physical nature of offshore operations and installation requirements are expected to be broadly similar across offshore basins. As long as the similarities of projects dominate the differences the reference class comparison is expected to serve as a useful baseline. If, on the other hand, the differences dominate development, then the reliability of the baseline cost is limited.
- There is no U.S. activity or project data to draw upon outside of hypothetical studies, and so the expertise, experience, and assumption set of the cost estimator will play a large role in the reliability of the estimates. Cost statistics from European projects can serve a useful role to baseline expected U.S. cost if properly adjusted and normalized to create a suitable class set.
- Small and diverse sample sets are best characterized by simple statistical measures. Standard deviations allow cost ranges to reflect the level of project uncertainty and scope and the beliefs of the user will dictate which range to select.

The top–down approach is also subject to a number of limitations. The primary obstacles include:

- European markets, government support, levels of competition, and marine vessel capability are different from U.S. markets, and if these differences dominate project development, European cost will be a biased statistic for U.S. projects.
- European project costs provide guidance on anticipated U.S. cost but do not translate directly and may be subject to escalation or decline factors.
- The top-down approach relies on historic data which may or may not reflect future realities and is unlikely to account for technical change and learning impacts.

Appendix A. Cost Adjustment Example

European costs are frequently reported in Euros or British pounds, but may also be reported in Norwegian, Danish and Swedish currency. Due to the variance in exchange rates over time, the method for converting European costs to U.S. dollars has important impacts on cost comparisons. Table 8.9 shows the adjusted costs (in 2010 U.S. dollars) for a hypothetical €500 million wind farm built between 2000 and 2009 using the inflate-first and exchange-first methods.

Using the inflate-first method, a €500 million wind farm is equivalent to $866 million if online in 2000 and $778 million if online in 2005. By contrast, using the exchange-first method, the value of money increases because of the weakening of the dollar against the euro over the period. The differences in the two methods are dramatic. Since all of the relevant currencies increase in value relative to the dollar, the inflate-first method yields more intuitive results.

References

1. Ozkan D, Duffey MR (2011) A framework for financial analysis of offshore wind energy. Wind Eng 35(3):267–288
2. Flyvbjerg B (2008) Curbing optimism bias and strategic misrepresentation in planning: reference class forecasting in practice. Eur Plan Stud 16(1):3–21
3. 4COffshore (2010), Database of offshore wind projects. Lowestoft
4. Hassan G (2009) UK offshore wind: lefting the right course. British Wind Energy Association, London
5. Musial W, Ram B (2010) Large scale offshore wind power in the United States: Assessment of opportunities and barriers. National Renewable Energy Laboratory. Golden, CO. NREL/TP-500-40745
6. Gerdes G, Tiedmann A, Zeelenberg S (2006) Case study: European offshore wind farms—a survey for the analysis of the experiences and lessons learnt. Dena, Groningen

7. Alpha Ventus (2010) Alpha Ventus. http://www.alpha-ventus.de/index.php?id=80. Accessed 26 Oct 2010
8. Fitzgerald J (2005) Wind farm off Wicklow opens. Irish Times, Dublin
9. European Investment Bank (2010) Bard I offshore wind farm. http://www.eib.org/projects/pipeline/2009/20090432.htm. Accessed 24 Oct 2010
10. BERR (2007) Offshore wind capital grants scheme. Barrow offshore wind farm first annual report. London. 09/P45. http://webarchive.nationalarchives.gov.uk/+/http://www.berr.gov.uk/files/file50163.pdf. Accessed 24 Oct 2010
11. Talisman Energy (2006) The Beatrice wind farm demonstrator—Fact sheet. http://www.beatricewind.co.uk/press/facts.asp. Accessed 26 Oct 2010
12. European Investment Bank (2010) Belwind: EIB finances Europe's largest offshore wind farm. http://www.eib.org/projects/news/eib-finances-europes-largest-offshore-wind-farm.htm. Accessed 24 Oct 2010
13. Belwind (2010) Facts and figures. http://www.belwind.eu/index.php?page=feiten&lang=en. Accessed 26 Oct 2010
14. Pepper L (2001) Monitoring and evaluation of Blyth offshore wind farm. Department of Trade and Industry. London. DTI URN 01/686
15. Lemming JK, Morthorst PE, Clausen NE (2007) Offshore wind power experiences, potential and key issues for deployment. Riso National Laboratory, Frederiksborgvej. Denmark. RISO-R-1673(EN)
16. European Investment Bank (2010) Global Tech I offshore wind farm. http://www.eib.org/projects/pipeline/2009/20090259.htm. Accessed 24 Oct 2010
17. Bradbury J (2010) Jumbo joins up Greater Gabbard. Offshore Media Group. http://www.offshore247.com/news/art.aspx?Id=16602. Accessed 24 Oct 2010
18. Dong Energy (2007) 2006 Annual report. http://www.dongenergy.com/EN/Investor/reports/Pages/annual%20reports2.aspx. Accessed 24 Oct 2010
19. Dong Energy (2008) 2007 Annual report. http://www.dongenergy.com/EN/Investor/reports/Pages/annual%20reports2.aspx. Accessed 24 Oct 2010
20. Dong Energy (2009) 2008 Annual report. http://www.dongenergy.com/EN/Investor/reports/Pages/annual%20reports2.aspx. Accessed 24 Oct 2010
21. Stromsta K (2010) Dong launches 172 MW Gunfleet Sands offshore wind project. Releftgenews.com. http://www.releftgenews.com/energy/wind/article217994.ece?print=true. Accessed 24 Oct 2010
22. Vattenfall (2010) Wind farms. Stockholm. http://www.vattenfall.com/en/wind-farms.htm. Accessed 24 Oct 2010
23. Ministry of Foreign Affairs of Denmark (2010) Dong Energy soon to inaugurate Horns Rev 2 offshore wind farm. http://www.investindk.com/visNyhed.asp?artikelID=22253. Accessed 24 Oct 2010
24. BERR (2007) Offshore wind capital grants scheme. Kentish Flats offshore wind farm 2nd annual report. London. 09/P46
25. Lundgren K (2009) Centrica to build $1.2 billion wind farm, team up with SocGen. Bloomberg. http://www.bloomberg.com/apps/news?pid=newsarchive&sid=ateP2AcRakFI. Accessed 24 Oct 2010
26. Mastiaux F (2010) E.ON Offshore wind energy factbook. E.ON. http://www.eon.com/en/downloads/EON_Offshore_Factbook_April_2010_EN.pdf. Accessed 24 Oct 2010
27. MPI (2009) Lynn and Inner Dowsing wind farm. http://www.mpi-offshore.com/workboats-projects/lynn-and-inner-dowsing-offshore-wind-farm. Accessed 24 Oct 2010
28. Sørensen HC, Hansen LK, Larsen JHM (2002) Middelgrunden 40 MW offshore wind farm Denmark-lessons learned. Renewable Realities—Offshore Wind Technologies. Orkney, UK
29. Carter M (2007) North Hoyle offshore wind farm: design and build. In: Proceedings of the Institution of Civil Engineers, Energy, vol 160. pp 21–29
30. NordzeeWind (2008) Offshore wind farm Egmond aan zee general report. Ijmuiden, Netherlands. OWEZ_R_141_20080215. http://www.noordzeewind.nl/files/Common/Data/OWEZ_R_141_20080215%20General%20Report.pdf. Accessed 24 Oct 2010

31. May J (2009) North Hoyle and Rhyl Flats offshore wind farms: Review of good practice in monitoring, construction and operation. Presentation to Scottish National Heritage, Inverness Scotland. 17 July 2009
32. European Investment Bank (2009) Sheringham Shoal offshore wind farm. http://www.eib.org/projects/pipeline/2009/20090481.htm. Accessed 24 Oct 2010
33. C-Power (2010) Investment. http://www.c-power.be/English/investment/index.html. Accessed 24 Oct 2010
34. Renewable energy focus (2010) Repower turbines to Thornton Bank offshore wind farm expansion. November 26. http://www.renewableenergyfocus.com/view/14232/repower-turbines-to-thornton-bank-offshore-wind-farm-expansion. Accessed 24 Aug 2011
35. IEA (2005) Offshore wind experiences. International Energy Agency. IEA Publications, Paris
36. Dong energy (2009) Dong Energy sells minority stake in Walney offshore wind farm. http://www.dongenergy.com/Walney/News/data/Pages/DONGEnergysellsminoritystakeinWalneyOffshoreWindFarm.aspx. Accessed 24 Oct 2010
37. Prysmian (2010) Prysmian will develop the power links for the Walney offshore wind farm, in the Irish sea. Prysmian Energy. Milan
38. Barthelmie R, Frandsen S, Morgan C, Henderson T, Sorensen HC, Garcia C, Lemonis G, Hendriks B, O'Gallachóir B (2001) State of the art and trends regarding offshore wind farm economics and financing. Presented at the EWEA Special Topic Conference, Brussels, December 10–12
39. EWEA (2009) Oceans of opportunity. European Wind Energy Association, Brussels
40. Tegen S (2010) Wind turbine manufacturers in the U.S.: Locations and local impacts. Windpower 2010 Conference, Dallas
41. BLS (2011) International labor comparisons. Bureau of Labor Statistics. http://www.bls.gov/fls/. Accessed 8 Aug 2011
42. MEPS (2011) Steel prices. http://www.meps.co.uk/world-price.htm. Accessed 8 Aug 2011
43. Maritime Journal (2002) Seacore secures North Hoyle wind farm contract. http://www.maritimejournal.com/features101/marine-civils/port,-harbour-and-marine-construction/seacore_secures_north_hoyle_wind_farm_contract. Accessed 24 Oct 2010
44. OilVoice (2008) Subocean announces another multi-million pound offshore windfarm contract. Oil Voice, September 17, 2008. http://www.oilvoice.com/PrinterFriendly/Subocean_Announces_Another_Multimillion_Pound_Offshore_Windfarm_Contract/87adb104.aspx. Accessed 24 Oct 2010
45. Master Marine (2009) StatoilHydro and Statkraft have awarded a major contract to Master Marine. http://www.master-marine.no/index.php?option=com_content&task=blogsection&id=1&Itemid=99. Accessed 24 Oct 2010
46. Morgan CA, Snodin HM, Scott NC (2003) Offshore wind: economics of scale, engineering resource and load factors. Department of Trade and Industry. Bristol, London
47. ODE (2007) Study of the costs of offshore wind generation. Department of Trade and Industry, Bristol, London
48. DTI (2004) The world offshore renewable energy report 2004–2008. Department of Trade and Industry, London. URN 04/393CD
49. Kühn M, BierboomsWAAM, van Bussel GJW, Ferguson MC, Göransson B, Cockerill TT, Harrison R, Harland LA, Vugts JH, Wiecherink R (1998) Opti-OWECS Final report vol. 0: Structural and economic optimisation of bottom-mounted offshore wind energy converters—executive summary. Delft University of Technology, Delft, The Netherlands. IW-98139R
50. Kooijman HJT, de Noord M, Volkers CH, Machielse LAH, Hagg F, Eecen PJ, Pierik JTG, Herman SA (2001) Cost and potential of offshore wind energy on the Dutch part of the North Sea. EWEA Special Topic Conference Offshore Wind Energy, Brussels. December 10–12
51. Schellstede H (2007) Wind power: wind farms of the Northern Gulf of Mexico. Paper presented at Offshore Technology Conference, Houston, 30 April–3 May 2007
52. Ernst and Young (2009) Cost of and financial support for offshore wind. Department of Energy and Climate Change. London, URN 09D/534

53. Coastal Point Energy (2010) Introduction to Coastal Point Energy. http://www.millergroupofcompanies.com/presentations/CoastalPointEnergyPPT3.pdf. Accessed 8 Aug 2011
54. Weiss J, Chang J (2010) Prefiled direct testimony of Jurgen Weiss, Ph.D. and Judy Chang. Massachusetts Department of Public Utilities DPU 10-54 Exhibit AG-JWJC-1, August 20, 2010

Chapter 9
Modeling Offshore Wind Installation Costs

9.1 System Description

9.1.1 User Input

The user is required to provide data on project characteristics, including system generation capacity, turbine capacity, distance to port, and distance to shore. The user is also required to select a primary installation vessel class from a list of alternatives and a method of installation across each stage. The user may provide additional information, if known, regarding contract and market conditions.

9.1.2 System Input

System data specifies vessel configurations, specifications and expected dayrates, unit time statistics per operational stage, and vessel spread requirements. System data is derived from data on existing wind farms in Europe, consultation with industry participants, activities performed in the offshore oil and gas industry, hypothetical models, and other sources. All system data enters the model in terms of their maximum, minimum, and average value.

9.1.3 Derived Data

Model-derived data apply empirical relations to compute the inner-array cable length, number of substations required, and the number of export cables. The model derives data based on statistical relationships computed from current project configurations.

M. J. Kaiser and B. F. Snyder, *Offshore Wind Energy Cost Modeling*,
Green Energy and Technology, DOI: 10.1007/978-1-4471-2488-7_9,

9.1.4 Model Output

Cost calculators are developed across each stage of installation: foundation, turbine, inner-array cable, export cable, substation, scour protection, and mobilization. Each stage of activity is broken down into tasks that must be completed, and time durations are assigned to each activity stage based on system data and user input. Time durations are on a per-unit basis and user input and derived data specify the number of units installed. For each stage of installation, the model computes the expected cost as the time required multiplied by the daily cost. Daily cost is the sum of the installation vessel dayrate and the spread dayrate. Dayrates and spread requirements are determined by user and system input. A contingency and weather factor may be incorporated in the estimation.

9.2 User Data

9.2.1 Project Characteristics

Offshore wind farms are described by many variables, but four variables in particular—nameplate capacity, turbine capacity, distance to port, and distance to shore—provide a first-order characterization of the system.

Nameplate Capacity
Wind farm capacity is the main determinant of capital expenditures and installation costs since the size and number of turbines determine work requirements and vessel selection. Generation capacity is correlated with the geographic area the turbines require to extract energy from the wind and the cable length to connect individual turbines. Generation capacity and distance to shore determines the need for offshore substations and the number of export cables.

Turbine Capacity
Turbine capacity for a given system capacity determines the number of turbines and foundations required. Turbine selection, water depth, soil type, and wind regime determine the foundation and vessel requirements of installation. Turbine capacity is related to rotor diameter which determines the minimum spacing requirements at the site and inner-array cable length.

Distance to Port
The distance between the staging area[1] and the wind farm determines the time for marine vessels to pick-up material and equipment at port and return to the work site. The number of trips varies with the installation stage, degree of pre-assembly,

---

[1] It is possible that more than one staging area will be utilized in fabrication, assembly and load out.

method of installation, and vessel spread. In turbine installation, it is common for the installation vessel to load the turbines at the staging area. For foundation installation, it is more common for the installation vessel to remain on site while a vessel spread provides logistical support. In cable installation, the manner in which cable is laid and location of the supply play dominant roles and distance to port is not expected to be a relevant descriptor of operational activity.

Distance to Shore
Distance to shore proxies the length of export cable required which is the primary factor in determining installation cost. Onshore transition type and environmentally sensitive areas will impact export cable routing and installation cost but are not considered because of unknown site-specific conditions.

Inner-Array Cable Length, Number of Offshore Substations
The length of the inner-array cable and number of offshore substations are positively correlated with the farm generation capacity and distance to shore. If known, they may be provided by the user; otherwise, the model derives expected values for these parameters based on average project characteristics.

9.2.2 Vessel Selection

The user selects a main vessel for each stage of installation from a list of possible alternatives. When the user selects the vessel type, the model provides transit speed, transport capacity, installation time, dayrate, and spread requirements. The same vessel can be used for foundation and turbine installation and cable laying or separate vessels can be used across each stage. The user can also create or modify an existing vessel by specifying their own parameters.

9.2.3 Installation Strategy

Foundation
Two strategies for foundation installation are considered, referred to as "self-transport" and "barge". In self-transport, the installation vessel will pick-up monopiles and transition pieces at the staging area, return to site, and begin operations. In barge installation, a barge will transport the monopiles and transition pieces from the staging area to the site and the installation vessel will be stationed at site and perform the required activities. We do not allow for monopiles to be wet-towed (floated) to site, and assume that the installation vessel will drive the pile and install the transition piece in sequence.

Turbine
All the turbine components will be transported on an installation vessel and assembled at site. The capacity of the vessel to transport turbines has a default value, but may be modified to account for onshore assembly and reductions in capacity. Barge transport is not considered.

Cable
For inner-array cable, the most common installation methods involve the use of an ROV operated by either the main installation vessel or a specialized cable-laying vessel. We model the latter option and assume inner-array cables are installed by a special-purpose vessel. For export cable, installation is carried out in a single continuous operation and transport to and from port is not required.[2]

9.2.4 Adjustment Factor

A user-defined adjustment factor is utilized to discount/escalate cost or activity duration to account for specific regional or market conditions. The adjustment factor is subjective and depends upon factors that are difficult to predict including scale economies, learning effects, project novelty, contract terms, and levels of competition. An adjustment factor of 1.5 indicates a 50% escalation, and an adjustment factor of 0.5 indicates a 50% discount. Model default value for the adjustment factor is unity. All else equal, we would expect small, near-term U.S. projects to have an adjustment factor greater than one due to a lack of scale economies, a slow learning curve, and constrained vessel supply. Application of novel techniques would also likely create an adjustment factor greater than one.

9.3 System Data

9.3.1 Vessel Specification

The user selects the vessel class from the available options or creates a new option. Each vessel type is specified by its transport capacity (number of foundations and transition pieces, number of turbine components, tons cable), speed (knots), expected dayrate ($/day) and installation time (day/foundation, day/transition piece, day/turbine, day/cable km) ranges, and expected spread requirements ($/day).

9.3.2 Expected Time

For each stage of activity, a unit time is assigned to perform the work. In foundation and turbine installation, the unit time is derived by model computation based on user selection of vessel type, installation strategy and distance to port,

[2] High-voltage export cable is expected to be imported from Europe or Asia by the installation vessel. Transport costs are borne by the developer and would normally be included as part of the installation contract.

and the system input of load time, inner-array movement time, and installation time. Time is expressed as hours per foundation and hours per turbine. In cable laying, unit time is expressed in days per km installed. For substations and scour protection, a fixed unit time is employed.

9.3.3 Vessel Dayrates

Each vessel type is associated with an expected dayrate so that selection of the vessel determines the installation dayrate. Vessel type and installation strategy determines the spread requirements and the expected spread dayrate.

9.4 Installation Stage Computations

9.4.1 Foundation and Turbine Installation

Foundation and turbine cost estimates are governed under a similar conceptual framework and are discussed together using the term "unit" to refer to either a foundation or turbine. For foundations, two models (self-transport and barge) are evaluated (Fig. 9.1). For turbines, only the self-transport model is considered; the barge model is rejected due to the risk associated with offshore lifts.[3] The self-transport models for turbines and foundations are identical in structure and differ only in parameterization.

Self-transport

In self-transport, the installation vessel picks up the components for one or more foundations or turbines, delivers them to the work site, and performs installation. Installation proceeds on a per trip basis. The total time per trip is the sum of the travel time, loading time, installation time, and intra-field movement time. When the vessel capacity is exhausted it returns to port for unit replenishment.

The total travel time (TRAVEL, h) is determined by the average speed of the vessel (S, kn) and the distance to port (D, nm):

$$\text{TRAVEL} = 2\left(\frac{D}{S}\right) \tag{9.1}$$

Total per-trip installation time (INSTALL, h), total loading time (LOAD, h) and total per-trip intra-field movement time (MOVE, h) are a function of vessel

[3] This reflects current industry practice. In cases where turbine components are supplied by a feeder vessel, the feeder was an elevating vessel.

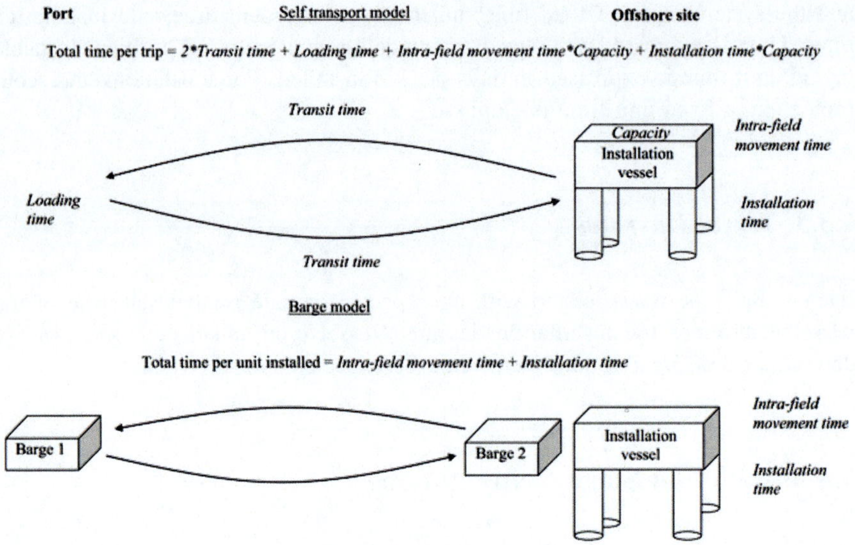

Fig. 9.1 Illustration of self-transport and barge models. Model parameters are in italics

capacity (VC, units/trip) and installation (I, h), loading (L, h) and intra-field movement (M, h) times per turbine:

$$\text{INSTALL} = \text{VC} \times I \tag{9.2}$$

$$\text{LOAD} = \text{VC} \times L \tag{9.3}$$

$$\text{MOVE} = \text{VC} \times M \tag{9.4}$$

The total time per trip (TRIP, h) is the sum of Eqs. (9.1)–(9.4):

$$\text{TRIP} = \text{TRAVEL} + \text{LOAD} + \text{INSTALL} + \text{MOVE} \tag{9.5}$$

The total time per trip is multiplied by a weather-adjustment factor (W) to account for the proportion of time vessels are able to operate. $W = 1$ indicates that there are no weather delays, while $W = 0.5$ indicates that half of the time vessels are unable to operate. The weather-adjusted time per trip (ADJTRIP, h) is:

$$\text{ADJTRIP} = \text{TRIP} \times \left(\frac{1}{W}\right) \tag{9.6}$$

The weather-adjusted time per trip divided by the vessel capacity yields the weather-adjusted installation time per unit (ADJUNIT, h):

$$\text{ADJUNIT} = \frac{\text{ADJTRIP}}{\text{VC}} \tag{9.7}$$

ADJUNIT is compared to known values in Appendix A to determine reasonableness of the assumptions and allows for comparison with the barge model but is not otherwise used in the model.

The number of units (NUMUNIT) is computed by the total farm capacity (FC, MW) divided by the turbine capacity (TC, MW):

$$\mathrm{NUMUNIT} = \frac{\mathrm{FC}}{\mathrm{TC}} \tag{9.8}$$

The number of trips (NUMTRIP) required is determined from the number of units and the vessel capacity:

$$\mathrm{NUMTRIP} = \frac{\mathrm{NUMUNIT}}{\mathrm{VC}} \tag{9.9}$$

The installation time (INTTIME, h) for the total operation is determined from:

$$\mathrm{INTTIME} = \mathrm{ADJTRIP} \times \mathrm{NUMTRIP} \tag{9.10}$$

Since the total cost of the project is determined by the total installation time and the daily cost, the total daily cost (TDC, \$) is the sum of the vessel dayrate (VDR, \$) and the spread dayrate (SDR, \$):

$$\mathrm{TDC} = \mathrm{SDR} + \mathrm{VDR} \tag{9.11}$$

The spread and vessel dayrates are determined from the system input and the user selection of vessel type and installation method (in this case, self-transport). The total cost (COST, \$) is the installation time normalized to days multiplied by Eq. (9.11):

$$\mathrm{COST} = \frac{\mathrm{INTTIME}}{24} \times \mathrm{TDC} \tag{9.12}$$

Cost may be multiplied by the adjustment factor, if desired, to reflect site- and time-specific variations that are not captured in the model framework (e.g., contract type, competition, and technology).

A summary of the model parameters, abbreviations and units is given in Table 9.1 and an example parameterization and calculation steps are provided in Table 9.2. The parameterization in Table 9.2 is hypothetical. In this case, the total time to install four turbines is approximately 600 h (25 days) and the total cost is approximately \$75 million.

Barge

The computational steps of the barge model are illustrated in Table 9.3. In the barge approach, it is assumed that the installation vessel is always supplied with units, so there is no downtime associated with vessel spread constraints.[4] Since the main installation vessel does not travel to and from port, travel time (TRAVEL)

[4] This is a realistic assumption as long as the vessel spread is adequate.

Table 9.1 Definition of terms used in foundation and turbine installation models

Variable name	Variable type	Models*	Description (unit)	Value
D	User	ST	Distance to port (nm)	
FC	User	ST, B	Farm capacity (MW)	
TC	User	ST, B	Turbine capacity (MW)	
VC	System	ST	Vessel capacity (number of units)	
LOAD	System	ST	Load time per trip (h)	
M	System	ST, B	Inner array move time per unit (h)	
S	System	ST, B	Vessel speed (knots)	
I	System	ST, B	Installation time per unit (h)	
L	System	ST	Load time per unit (h)	
W	System	ST, B	Weather-adjustment factor (%)	
VDR	System	ST, B	Main installation vessel dayrate ($/day)	
SDR	System	ST, B	Total spread dayrate ($/day)	
TDC	Computation	ST, B	Total daily cost ($)	VDR + SDR
INSTALL	Computation	ST	Installation time per trip (h)	VC*I
MOVE	Computation	ST	Inner-array move time per trip (h)	VC*M
TRAVEL	Computation	ST	Travel time per trip (h)	2*(D/S)
TRIP	Computation	ST	Total time per trip (h)	TRAVEL + LOAD + INSTALL + MOVE
ADJTRIP	Computation	ST	Weather-adjusted time per trip (h)	TRIP/W
NUMUNIT	Computation	ST, B	Number of units in the farm	FC/TC
NUMTRIP	Computation	ST	Number of trips required	NUMUNIT/VC
INTTIME	Computation	ST	Installation time for the total farm (h)	ADJTRIP*NUMTRIP
COST	Computation	ST, B	Total cost of installation ($)	(INTTIME/24)*TDC
UNIT	Computation	B	Total time to install one unit and move to next unit (h)	I + M
ADJUNIT	Computation	ST, B	Weather-adjusted time to install one unit and move (h)	UNIT/W, (B) or ADJTRIP/VC, (ST)
INTTIME	Computation	B	Installation time for the total farm (h)	ADJUNIT*NUMUNIT

Note * ST denotes self-transport; B denotes barge

and loading time (LOAD) are not required and the model changes from calculating times per trip to time per unit.

The time per unit (UNIT, h) is the sum of the time to install a unit (I, h) and move to another intra-field location (M, h):

$$\text{UNIT} = I + M \tag{9.13}$$

The weather-adjusted time per unit is Eq. (9.13) divided by W:

$$\text{ADJUNIT} = \text{UNIT} \times \left(\frac{1}{W}\right) \tag{9.14}$$

The total installation time is the product of the weather-adjusted time per unit and the number of units installed:

$$\text{INTTIME} = \text{ADJUNIT} \times \text{NUMUNIT} \tag{9.15}$$

Table 9.2 Example parameterization and calculation steps of the self-transport model

Parameters	Step	Computation method	Example
$L = 3$ h $I = 96$ h $M = 8$ h VC = 4 $W = 75\%$ $S = 10$ knots $D = 100$ nm VDR = \$100,000 SDR = \$20,000 FC = 300 MW TC = 3 MW	Determine travel time per trip	$\text{TRAVEL} = 2\left(\frac{D}{S}\right)$	$\text{TRAVEL} = 2 \times \frac{100}{10} = 20\,\text{h}$
	Determine installation time per trip	$\text{INSTALL} = \text{VC} \times I$	$\text{INSTALL} = 96 \times 4 = 384\,\text{h}$
	Determine movement time per trip	$\text{MOVE} = \text{VC} \times \text{M}$	$\text{MOVE} = 8 \times 4 = 32h$
	Determine loading time per trip	$\text{LOAD} = \text{VC} \times \text{L}$	$\text{LOAD} = 3 \times 4 = 12h$
	Determine total time per trip	$\text{TRIP} = \text{TRAVEL} + \text{LOAD} + \text{INSTALL} + \text{MOVE}$	$\text{TRIP} = 12 + 20 + 384 + 32 = 448\,\text{h}$
	Adjust for weather	$\text{ADJTRIP} = \text{TRIP} \times \left(\frac{1}{W}\right)$	$\text{ADJTRIP} = \frac{448}{0.75} = 597.3\,\text{h}$
	Determine number of units required	$\text{NUMUNIT} = \frac{\text{FC}}{\text{TC}}$	$\text{NUMUNIT} = \frac{300}{3} = 100$
	Determine number of trips required	$\text{NUMTRIP} = \frac{\text{NUMUNIT}}{\text{VC}}$	$\text{NUMTRIPS} = \frac{100}{4} = 25$
	Determine total time required	$\text{INTTIME} = \text{ADJTRIP} \times \text{NUMTRIP}$	$\text{INTTIME} = 597.3 \times 25 = 14933\,\text{h}$
	Determine total daily cost	$\text{TDC} = \text{SDR} + \text{VDR}$	$\text{TDC} = 20{,}000 + 100{,}000 = \$120{,}000$
	Determine total cost	$\text{COST} = \frac{\text{INTTIME}}{24} \times \text{TDC}$	$\text{COST} = \frac{14{,}933}{24} \times 120{,}000 = \$74{,}665{,}000$

Table 9.3 Example parameterization and calculation steps of the barge model

Parameters	Step	Computation method	Example
$I = 96$ h	Determine total time per unit	UNIT = I+M	UNIT = 94 + 8 = 104 h
$M = 8$ h	Adjust for weather	$\text{ADJUNIT} = \text{UNIT} \times \left(\frac{1}{W}\right)$	$\text{ADJUNIT} = \frac{104}{0.75} = 138.6\,\text{h}$
$W = 75\%$			
VDR = \$100,000	Determine no. of units required	$\text{NUMUNIT} = \frac{\text{FC}}{\text{TC}}$	$\text{NUMUNIT} = \frac{300}{3} = 100$
SDR = \$20,000			
FC = 300 MW	Determine total time required	INTTIME = ADJTRIP × NUMTRIP	INTTIME = 138.6 × 100 = 13, 860 h
TC = 3 MW	Determine total daily cost	TDC = SDR + UDR = \$120, 000	TDC = 20,000 + 100,000 = \$120,000
	Determine total cost	$\text{COST} = \frac{\text{INTTIME}}{24} \times \text{TDC}$	$\text{COST} = \frac{13.860}{24} \times 120,000 = \$69,300,000$

As before, the total cost is determined from the product of the time and the daily cost:

$$\text{COST} = \frac{\text{INTTIME}}{24} \times \text{TDC} \tag{9.16}$$

9.4.2 Cable Installation

Inner-array cables

Inner-array cable installation cost (ARRAYCOST, $) is determined by the vessel dayrate (ARRAYDR, $) and the time required[5] (ARRAYTIME, day):

$$\text{ARRAYCOST} = \text{ARRAYDR} \times \text{ARRAYTIME} \tag{9.17}$$

ARRAYTIME is determined from the cable length (ARRAYLENGTH, km) divided by the rate (ARRAYRATE, km/day) of installation:

$$\text{ARRAYTIME} = \frac{\text{ARRAYLENGTH}}{\text{ARRAYRATE}} \tag{9.18}$$

Inner-array cable length can be determined through various empirical relations related to rotor diameter and other variables. The relation adopted here was developed from installed European wind farms and describes cable length in terms of farm capacity (FC, MW):

$$\text{ARRAYLENGTH} = 0.00067(\text{FC})^2 + 14.6 \tag{9.19}$$

Export cables

Export cable installation cost (EXPORTCOST, $) is a function of the installation time (EXPORTTIME, day) and vessel dayrate (EXPORTDR, $):

$$\text{EXPORTCOST} = \text{EXPORTTIME} \times \text{EXPORTDR} \tag{9.20}$$

EXPORTTIME is set by the cable length (EXPORTLENGTH, km) and the installation rate (EXPORTRATE, km/day):

$$\text{EXPORTTIME} = \frac{\text{EXPORTLENGTH}}{\text{EXPORTRATE}} \tag{9.21}$$

---

[5] A model of installation time involving travel time was not considered because the data available for parameterization already included transit times. Adding transit time to the model would double-count.

9.4.3 Substation Installation

Substation installation is composed of foundation installation and topside installation. We assume that a jacket foundation will be barged to site and placed with a heavy-lift vessel. Pile driving operations secure the jacket to the seabed and represent the primary time in jacket installation. Driving time depends on the soil type, the depth of burial, the size and thickness of the piles, and the number of piles. For topside installation, a heavy-lift vessel is assumed to work for three days (see Chap. 5, Sect. 5.5). The total cost is the sum of the foundation and topside installation cost.

9.4.4 Scour Protection

Scour protection is installed by a barge/tug traveling between port and each turbine site. The total time per trip is composed of loading rock, travel, and installation (dumping) time. Inner-array movement time is ignored. The total tonnage of scour required (TSR, ton) is given by the tonnage of scour per unit (SPU, ton) times the number of units (NUMUNIT):

$$\mathrm{TSR} = \mathrm{SPU} + \mathrm{NUMUNIT} \tag{9.22}$$

The number of trips required (SCOURNUMTRIP) is the total scour requirements described by Eq. (9.22) divided by the vessel capacity (SCOURVC, ton)

$$\mathrm{SCOURNUMTRIP} = \frac{\mathrm{TSR}}{\mathrm{SCOURVC}} \tag{9.23}$$

9.4.5 Mobilization

Models for mobilization costs are readily developed, but since mobilization distances are unlikely to be known until the late planning stages of a project, the user may not feel confident specifying a value. A default value of 1,000 nm is suggested. In either case, a flat-rate based on the average vessel class-specific mobilization cost is added to the total costs. We assume that after project completion the vessel returns to port and demobilization costs are equal to mobilization costs. Spread vessels will be locally supplied in most situations and are assumed to not require mobilization costs.

Table 9.4 Turbine and foundation installation vessel parameters

Vessel type	Speed (kn)	Foundation capacity	Turbine capacity	Dayrate range ($/day)	Expected dayrate ($/day)
Liftboat	4–6	0	1–2	12,500–75,000	35,400
JU Barge	4–8	2–4	2–6	25,000–150,000	64,200
SPIV	8–12	4–8	6–8	60,000–300,000	134,300

Table 9.5 Total spread dayrate costs by vessel category and transport system

		Number of vessels			Total dayrate*		
Vessel	Transport system	Tugs	Barges	Crewboats	Min	Max	Average
SPIV	Barge	2–3	2–3	2–4	18,500	30,250	24,375
SPIV	Self-transport	0	0	1–3	2,500	7,500	5,000
Jackup	Barge	3–4	2–3	2–4	23,500	35,250	29,375
Jackup	Self-transport	1–2	0	1–3	7,500	17,500	12,500
Liftboat	Barge	2–3	2–3	2–4	18,500	30,250	24,375
Liftboat	Self-transport	0–1	0	1-3	2,500	12,500	7,500

Note * Dayrates of tugs, barges, and crewboats are assumed to be $10,000, $1,750, and $3,500 per day, respectively

9.5 Model Parameterization

9.5.1 Vessel Data

Three vessel types are available: a liftboat, a jackup barge, or a self-propelled installation vessel (SPIV). In Table 9.4, the installation vessel parameters are based on the analysis described in Chaps. 6 and 7. When the installation vessel is selected, the vessel speed, capacity, and dayrates are input into the model. For speed and capacity, the expected value is the average of the range. Table 9.5 summarizes input on spread requirements and costs. The dayrates of tugs, barges, and crewboats are assumed to be 10,000, 1,750, and $3,500/day, respectively. Spread dayrates were not given a range because the number of vessels in the spread varies. Vessel spreads are not expected to differ substantially for foundations and turbines.

9.5.2 Foundations

Table 9.6 depicts the mean and range of system inputs for loading time, installation time, intra-field movement time, and weather uptime. The largest single temporal variable in the foundation parameterization is installation time. Installation time consists of the time to drive a pile plus the time to place the transition piece over the monopile, level and grout in place. Pile driving time is estimated

Table 9.6 Parameterization range for factors influencing foundation installation time

Model	Load time, L (h)	Installation time, I (h)	Movement time, M (h)	Weather uptime, W (%)
Self-transport	2–4 (3)	36–96 (72)	4–8 (6)	75–95 (90)
Barge	NA	36–96 (72)	NA	75–95 (90)

Note Expected values depicted in parentheses

Table 9.7 Foundation installation time as a function of turbine capacity

Turbine capacity (MW)	Installation time range, I (h)	Expected value, I (h)
2.5	36–48	40
3	36–72	54
3.6	48–72	60
4	72–96	84
5	96	96

based on industry time tables. The depth at which piles are driven is site-specific and depends upon water depth, soil type, loading requirements, and environmental conditions. Most piles are driven 80–150 ft into the seabed. Driving rates for 150–200 in. (4–5 m) diameter and 1.5–3 in. (4–8 cm) wall thickness range from 0.2 to 0.4 h/ft. Total pile driving times may thus range from 16 to 60 h. Time is also required for transition piece placement and grouting. A total installation time range of 36–96 h is considered reasonable in most cases. Wall thickness, pile diameter, and insertion depth are determined in part by turbine capacity, and the model is parameterized with short installation times for smaller turbines and longer installation times for larger capacity turbines (Table 9.7). Appendix A verifies the parameterization.

9.5.3 Turbines

System inputs for loading time, installation time, intra-field movement time, and weather uptime for turbines are shown in Table 9.8. Installation time and vessel capacity may vary with turbine electrical capacity. Increases in turbine capacity will decrease the number of turbines that may be carried per trip and may increase the number of lifts required. The interaction between turbine size, installation time, and vessel choice is shown in Table 9.9. Based on the relationship between vessel size and crane capacity, larger, more expensive vessels are assumed to install turbines in fewer lifts and less time than smaller vessels; we assume that a SPIV can install a turbine in half the time required by a liftboat, with a jackup barge approximately intermediate between the two options.[6]

[6] As the number of projects and vessels increases, data on the installation rate by vessel class may be available to test this assumption.

Table 9.8 Parameterization range for factors influencing turbine installation time

Model	Load time, L (h)	Installation time, I (h)	Movement time, M (h)	Weather uptime, W (%)
Self-transport	2–6 (4)	36–120 (72)	4–8 (6)	75–90 (85)

Note Expected values depicted in parentheses

Table 9.9 Turbine installation time by capacity and vessel type

Turbine capacity (MW)	Vessel type	Installation time, I (h)	Expected value, I (h)
2.5–3	LB	72–96	84
	JU	48–72	60
	SPIV	36–48	42
3–4	LB	96–120	108
	JU	60–96	72
	SPIV	48–60	54
4–5	LB	NA	NA
	JU	72–120	96
	SPIV	60–96	72

9.5.4 Cables

Inner-array

Inner-array cables are assumed to be installed by either a dynamically positioned self-propelled vessel or a barge-tug system. A medium sized vessel such as a modified offshore supply vessel or a barge-tug spread is expected to have a dayrate of $25,000/day while a larger vessel is expected to have a dayrate of $75,000/day. Without additional information on vessel type, the model assumes an expected dayrate of $50,000. Installation rate data derived from offshore projects specifies a lay rate of 0.3 km/day with a range of 0.15–0.60 km/day.

Export

High-voltage export cables are heavy and require a large vessel equipped with a high capacity turntable. These vessels are expected to cost approximately $125,000/day with a range between 75,000 and 175,000 $/day [1]. Empirically derived export cable installation rates range from 0.2 to 1.4 km/day with an expected value of 0.7 km/day. For a $125,000 per day vessel laying at 0.7 km/day, the cost of high-voltage cable installation is $178,571 per km.[7] For medium-voltage export cable, the installation time does not change, but the dayrate is expected to be lower. The number of export cables used in existing wind farms is

[7] This compares reasonably well with other estimates. Recently, De Alegria et al. [2] estimated installation costs for HVAC cable as 100,000 €/km (approximately 134,000 $/km in 2009 dollars). Other estimates range from 200,000 to 750,000 $/km [3–5].

Table 9.10 Number of export cables and substations required by distance to shore and generation capacity

		Capacity (MW)			
	Distance (km)	<100	100–200	200–300	>300
Export Cables	≤10	2*	1-3*	1	2
	11–20	1	1	1	2
	21–30	1	1	1	2
	>30	1	1	2	2
Substations	≤10	0	0–1	1	1
	11–20	1	1	1	2
	21–30	1	1	1	2
	>30	1	1	1	2

Note * Indicates medium-voltage cable

shown in Table 9.10. Cable length is estimated by distance to shore, however, this proxy can be considered a lower bound because of the potential need to avoid sensitive areas or connect to existing onshore infrastructure.

9.5.5 Substation

Substation foundations are expected to be barged to site; therefore, installation time is simply the time to place the jacket and drive piling, and to apply grout, if required. Pile driving is expected to range between 0.02 and 0.05 h/foot, assuming pile diameters between 18 and 48 in. and wall thickness of 1–2 inches. The total pile driving time for 4–6 piles, each driven 100–150 ft into the seabed, will range between 8 and 45 h with additional time required to reposition the vessel between piles (assumed to be 8 h total). Most shallow water jackets will weigh between 500 and 1500 tons and require approximately 20 h to place. Total foundation installation time is expected to range between 1.5 and 3 days. Substation topsides are assumed to require three days to install.

We assume a heavy-lift vessel is required for foundation and topside lifting. Heavy-lift vessels are assumed to have dayrates of $80,000/day (range 50,000–$115,000/day) for lifts under 1,000 tons, and $150,000/day (range 140,000–$500,000/day) for lifts over 1,000 tons. For a lift of unknown weight, we assume $100,000 per day. Additionally, a spread consisting of three tugs, one barge and one crewboat is required at a cost of $19,200 per day. The number of substations required is estimated from Table 9.10 based on the distance to shore and farm capacity.

9.5.6 Scour Protection

Scour protection is site-specific. In some cases, scour protection may not be needed, while at other locations the amount of protection will vary with the sediment and current conditions [6]. When needed, the total mass of scour

Table 9.11 Ranges of mobilization costs by mobilization distance and vessel type (1000 $)

	Liftboat		Jackup		SPIV		Heavy-lift	
Distance (nm)	Exp	Range	Exp	Range	Exp	Range	Exp	Range
250	129	104–177	504	205–801	192	151–234	273	269–276
500	246	171–353	698	335–1,073	385	302–467	522	476–549
1,000	479	304–707	1,086	595–1,618	769	604–934	1,018	877–1098
1,500	712	437–1,060	1,473	855–2,162	1,154	907–1,402	1,515	1,278–1,647
2,000	1,061	570–1,413	1,861	1115–2,707	1,539	1,209–1,869	2,011	1,679–2,196

Note Exp denotes expected value

protection required is assumed to be approximately 1,250 tons per turbine with a range of 1,000–1,500 tons [7]. We assume that a tug and hopper barge traveling at 4 knots with a 1250 tons capacity is leased for $8,000 per day. Dumping is assumed to take 4 h and loading 12 h per trip.

9.5.7 Mobilization

Mobilization costs depend on mobilization distances, vessel type, and vessel dayrates. The user may select any mobilization cost from Table 9.11 or the default value of 1,000 nm. If dayrates other than the expected dayrates are used, the mobilization cost may be adjusted proportionally.

9.6 Hypothetical Example

9.6.1 Set-Up

We illustrate the model using a hypothetical 300 MW wind farm composed of 83, 3.6 MW turbines, 100 nm (185 km) from port and 10 nm (19 km) from shore. Installation vessels are not specified to allow comparisons. Vessel dayrates are fixed at expected values. Cost estimates are depicted per stage of installation.

9.6.2 Component Costs

Table 9.12 summarizes the cost estimate range for foundation and turbine installation. Jackups and SPIVs are examined using both the barge and self-transport model. In each case, expected values for the vessel speed, capacity, dayrates, and spread dayrates are employed. Foundation costs are similar between the barge and self-transport models but vary by vessel type with jackup barges significantly less expensive than SPIVs. Within a model class and vessel type, the

Table 9.12 Hypothetical 300 MW wind farm—foundation and turbine installation cost

	Model	Vessel type	Load time, L (h)	Installation time, I (h)	Movement time, M (h)	Weather factor, W (%)	Cost (million $)
Foundation	Barge	JU	NA	48	6	90	23.8
	Barge	JU	NA	60	6	90	29.0
	Barge	JU	NA	72	8	75	42.3
	Barge	SPIV	NA	48	6	90	36.3
	Barge	SPIV	NA	60	6	90	44.4
	Barge	SPIV	NA	72	8	75	64.5
	Self-transport	JU	3	48	6	90	21.0
	Self-transport	JU	3	60	6	90	25.0
	Self-transport	JU	4	72	8	75	35.6
	Self-transport	SPIV	2	48	6	90	31.2
	Self-transport	SPIV	3	60	6	90	37.7
	Self-transport	SPIV	3	72	8	75	54.5
Turbine	Self-transport	LB	2	96	4	90	23.4
	Self-transport	JU	2	60	4	90	24.2
	Self-transport	SPIV	2	48	4	90	30.1
	Self-transport	LB	4	108	6	85	27.8
	Self-transport	JU	4	72	6	85	30.8
	Self-transport	SPIV	4	54	6	85	36.6
	Self-transport	LB	6	120	8	75	34.8
	Self-transport	JU	6	96	8	75	45.4
	Self-transport	SPIV	6	60	8	75	46.9

minimum expected cost is approximately 60% of the maximum cost suggesting that variation in activity duration can cause significant uncertainty in cost.

A 300 MW wind farm is expected to require approximately 75 km of inner-array cables, and based on its distance from shore (19 km) is expected to require one export cable and substation (from Table 9.10). Vessel dayrate is constant at $50,000/day for inner-array cable placement and $125,000/day for export cables. Given these conditions, Table 9.13 shows cost estimates for cable installation for variable lay rate. Inner-array cable costs more than export cable, and neither cable installation is as expensive as turbine or foundation installation. Substation installation and scour protection costs are small relative to the total cost.

9.6.3 Total Costs

Table 9.14 shows the expected component costs, the expected total cost, and the costs per stage as a proportion of the total. If the total capital costs for a 300 MW wind farm constructed in 2012–2017 are approximately $1.3 billion (assuming 4.5 million $/MW), the installation costs are a relatively small proportion of the total costs, approximately 7% in the expected case and up to 11% in the maximum scenario.

Table 9.13 Hypothetical 300 MW windfarm—cable, substation, scour protection costs

Component	Parameterization	Lay rate (km/day)	Dayrate ($/day)	Distance (km)	Cost (million $)
Inner-array	Minimum	0.6	50,000	74.9	6.2
	Expected	0.3	50,000	74.9	12.5
	Maximum	0.15	50,000	74.9	24.0
Export	Minimum	1.4	125,000	37	3.3
	Expected	0.7	125,000	37	6.6
	Maximum	0.2	125,000	37	23.1
		Duration (days)	Dayrate ($/day)	Spread ($/day)	Cost (million $)
Substation	Minimum	4.2	80,000	19,200	0.42
	Expected	5.0	100,000	19,200	0.60
	Maximum	5.7	150,000	19,200	0.96
		Number of trips	Dayrate ($/day)	Trip time (days)	Cost (million $)
Scour	Minimum	69	8,000	2.75	1.5
	Expected	83	8,000	2.75	1.8
	Maximum	104	8,000	2.75	2.3

Table 9.14 Hypothetical 300 MW wind farm—total costs

	Parameterization	Foundation	Turbine	Inner array	Export	Sub-station	Scour	Mob	Total
Cost (million $)	Minimum	23.8	30.1	6.2	3.3	0.4	1.5	2.9	68.2
	Expected	29.0	36.6	12.5	6.6	0.6	1.8	2.9	90
	Maximum	42.3	46.9	25.0	23.1	1.0	2.3	2.9	143.5
Percentage (%)	Minimum	34.9	44.1	9.1	4.8	0.6	2.2	4.3	100
	Expected	32.2	40.7	13.9	7.3	0.7	2.0	3.2	100
	Maximum	29.5	32.7	17.4	16.1	0.7	1.6	2.0	100

9.6.4 Sensitivity Analysis

The sensitivity of the self-transport model to changes in farm capacity, distance to port, turbine capacity, installation time, and dayrates is depicted in Figs 9.2, 9.3, and 9.4. Parameters not otherwise varied are set at their expected values and the costs represent only the output of the self-transport model for turbine installation.

The model scales linearly with farm capacity, dayrate, installation time, and distance to port, but nonlinearly with changes in turbine capacity. For every 1 MW increase in farm capacity, the installation cost increases approximately $132,000; for every nautical mile further from port, the cost increases $16,000; for every one thousand dollar increase in dayrate, the cost increases by $283,000, and for every 1 h increase in installation time per turbine, project costs increase by $569,000. The linearity of the relationship between capacity and cost suggests that

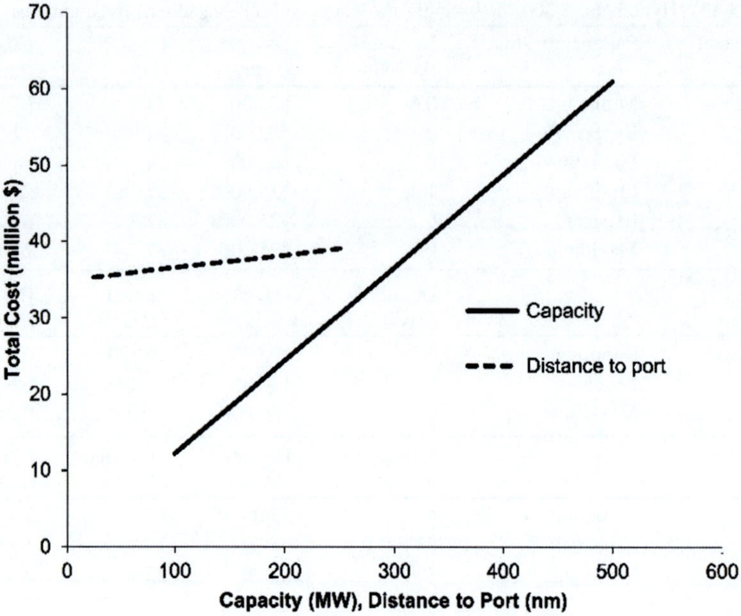

Fig. 9.2 Sensitivity of the self-transport model to changes in capacity and distance to port

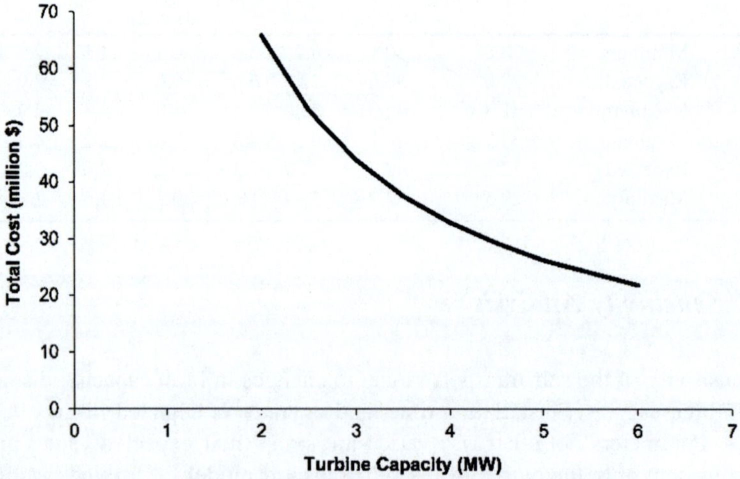

Fig. 9.3 Sensitivity of the self-transport model to changes in turbine capacity

economies of scale do not exist relative to total farm capacity, but the relationship in Fig. 9.3 suggests that there are significant economies associated with increasing turbine size. The sensitivity of the self-transport model to changes in dayrate and installation time is depicted in Fig. 9.4.

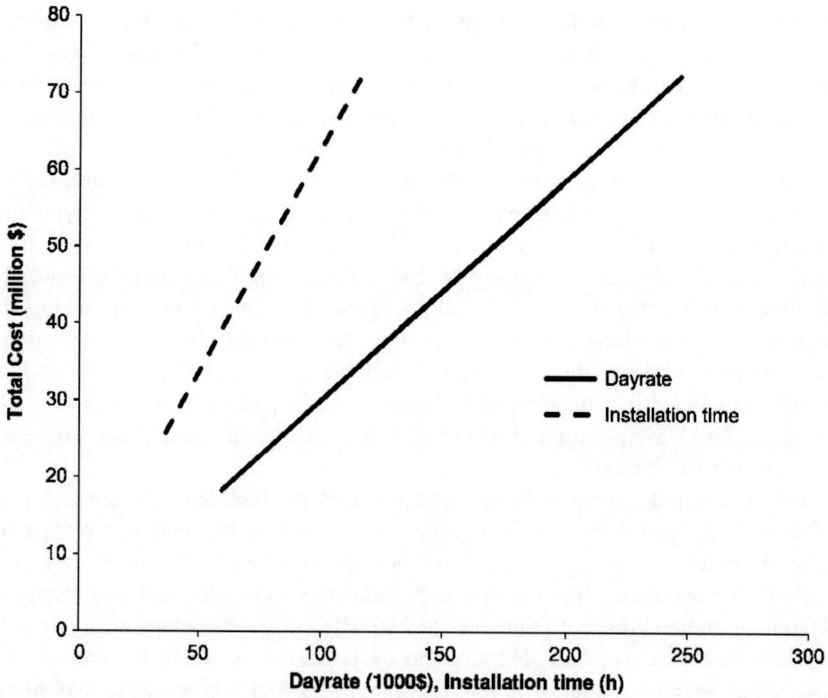

Fig. 9.4 Sensitivity of the self-transport model to changes in dayrate and installation time

9.7 Proposed U.S. Projects

9.7.1 Cape Wind: Massachusetts

Cape Wind is an offshore project proposed for the south side of Cape Cod and is likely to be one of the first offshore wind developments in the United States. The U.S. Department of the Interior has given regulatory approval to project developers, but construction activities have not begun as of December 2011. A detailed Environmental Impact Statement was prepared for the Cape Wind project [8] which allows for a detailed parameterization.

The site selected for Cape Wind is approximately 6 nm from land and will utilize 130, 3.6 MW turbines on monopile foundations with a total nameplate capacity of 468 MW. Foundations are to be installed with a jackup barge using the barge method; turbines are to be installed by a SPIV. Quonset, Rhode Island is the designated port and is approximately 60 nm to the proposed site.

The EIS contains additional information beyond the required user input which is incorporated into the model. The EIS states the turbine erection is expected to take between 30 and 40 h per unit (installation time plus inner-array movement time) and that the SPIV is expected to carry 6–8 turbines (with transition pieces) per trip. The time to install a monopile is less certain, but the developers state that the monopiles will be installed over an 8-month period by two jackup barges. This gives a total time of 480 days and an installation time per monopile (installation time plus inner-array movement time) of approximately 3.7 days (89 h). Four to six spread vessels are expected to support foundation installation.

For cable installation the EIS estimates that 430 days will be required for the inner array and 19 days for the export cable. Based on the model parameterization, we would expect installation time to be 537 and 14 days for inner-array and export cables, respectively, but we employ the EIS values. Cape Wind engineers estimate the time to install the substation at 30 days, but we suspect this value includes a large amount of finishing work (welding on ladders, making electrical connections, etc.) that do not require the presence of a heavy-lift vessel. Therefore, we employ model estimates for substation installation time. For scour protection, the EIS estimates three days per turbine. One SPIV, two jackups and one heavy-lift vessel are expected to be mobilized.

The user input and additional input parameters from the EIS are given in Table 9.15. Not all entries are complete because not all parameters are used for each installation stage; all parameters not specified are set at default levels. Table 9.15 gives two output costs: one using the EIS input and one using all default parameters. For both outputs, the estimated cost was determined using the expected installation vessel dayrate; the cost range was found by varying the installation vessel dayrate to its minimum and maximum value while leaving all time-related terms constant. Therefore, the maximum and minimum values do not have as great a range as they would if both temporal and dayrate variables were allowed to vary.

The total cost from the model using the EIS derived input matches closely with the default input. The EIS-based parameterization output cost ranges from 68 to $222 million, with an expected cost of $121 million. These values are slightly lower than the model default cost which ranges from 74 to $254 million with an expected cost of $136 million. The largest cost difference between the two models was in turbine installation costs. Foundations and turbine installation is expected to comprise between 70 and 80% of total installation cost, followed by inner-array cable which accounts for 15–19% of the total cost. If Cape Wind is developed at a capital cost of 4.5 million $/MW, the development cost will be approximately $2.1 billion. For this unit cost, installation expenditures will range between 3 and 11% of capital costs.

9.7.2 Bluewater Wind: Delaware

Bluewater Wind plans to build a 450 MW wind farm 13 nm off the coast of Delaware, and as of October 2011, is awaiting regulatory approval and financing. Preliminary planning suggests a development using 3 MW turbines, two substations, and 4 export cables [9, 10]. No information on vessel class, supply strategy or port location has been released, but Bluewater has attempted to secure financing

Table 9.15 Model parameters and output cost estimates at Cape Wind

		Foundations	Turbines	Inner-array	Export	Substation	Scour	Mob	Total
User input	Distance to port (nm)		60				60		
	Distance to shore (nm)				6				
	Turbine capacity (MW)	3.6	3.6				3.6		
	Farm capacity (MW)	468	468	468			468		
	Vessel type	JU	SPIV						
	Transport system	B	ST						
System input	Spread size	Minimum	Expected						
	Installation time (h/unit)	84	31						
	Installation time (day)			430	19		390		
	Vessel capacity		7						
	Movement time (h)	5	4						
Output (million $)	Expected cost	56.0	33.6	21.5	2.4	0.6	3.1	4.0	121.2
	Maximum cost	102.0	72.9	32.3	3.3	1.0	3.1	7.3	221.9
	Minimum cost	35.0	15.9	10.8	1.4	0.4	3.1	1.5	68.1
Default (million $)	Expected cost	45.3	56.1	26.9	1.9	0.6	1.4	4.0	136.2
	Maximum cost	79.4	121.9	40.3	2.7	1.0	1.7	7.3	254.3
	Minimum cost	29.7	26.6	13.4	1.2	0.4	1.1	1.5	73.9

Table 9.16 Installation cost estimate at Bluewater Wind and Coastal Point wind farms (million $)

	Bluewater Wind (Delaware)			Coastal Point (Galveston)		
	Expected	Maximum	Minimum	Expected	Maximum	Minimum
Foundations	66.1	135.2	35.2	7.6	12.7	4.7
Turbines	65.1	142.2	30.3	12.2	23.5	5.7
Inner-array	25	37.6	12.5	4.9	7.4	2.5
Export	4.3	6	2.6	2.6	3.7	1.6
Substation	1.2	2	0.8	0.6	1	0.4
Scour	2.1	2.5	1.6	0	0	0
Mob	1.8	3.6	0.5	0	0	0
Total	165.6	329.1	83.5	27.9	48.3	14.9

to build its own SPIV [11] and we assume that an SPIV is the preferred choice for foundation and turbine installation. We assume that the project would use a port facility near Wilmington, DE, approximately 100 nm from the site, and that the barge method is used to supply foundations. All temporal parameters are set to their expected values, and dayrates are varied between their minimum and maximum levels.

The model output is shown in Table 9.16. The model output installation cost ranges from 84 to $329 million, with an expected cost of $166 million. Foundation and turbine installation is expected to comprise 78–85% of total cost; inner-array cable is expected to represent 11–15% of total cost. Assuming a capital expenditure of $4.5 million per MW, the installation costs are approximately 8% of the total cost.

9.7.3 Coastal Point Energy-Galveston: Texas

The Galveston Offshore Wind Project is under development by Coastal Point Energy (formerly WEST) and is planned for Texas state waters approximately 8 nm from shore. The wind farm is designed for 60, 2.5 MW turbines with a total capacity of 150 MW. Tripod foundations are specified in development plans. Coastal Point has also indicated they will employ a liftboat with a 500 ton lift capacity for installation. Public documents indicate that the developers expect foundations to require 1.5 days each, consistent with the model estimates for small turbines. We assume that the barge method is used for foundation transport and no mobilization costs are required as liftboat and heavy-lift vessels are common in the Gulf of Mexico. We also assume that scour protection is not required because of the nature of the foundations.

Model output is shown in Table 9.16 and the results match relatively well with estimates given by the developers [12]. The developers estimated the costs of foundation and turbine installation as $24.5 million; the expected model output is $19.8 million. The developers expect a total capital cost of $360 million ($2.4 million per MW). At $2.4 million per MW, the expected cost is 8% of the total costs; at \$4.5 million/MW, the maximum cost is about 7% of total cost. However, the expected costs assume a vessel dayrate similar to that required for an existing large liftboat. Coastal Point has indicated they may build a larger liftboat which will result in costs more similar to the maximum parameterization [13].

9.8 Model Limitations

Offshore wind development in the U.S. has not yet progressed to the construction stage, so empirical analysis on actual projects cannot be performed. To estimate offshore wind installation costs, it is necessary to develop models based on the requirements of the task and the expected time to perform and vessels needed. A large number of factors and events impact offshore construction and by selecting a small set of variables, only a portion of the system complexity is captured. However, the expectation is that the variables selected adequately describe the system and proxy the influence of unobservable variables.

There is uncertainty in the duration of work requirements and vessel dayrates and uncertainty in our ability to accurately predict these data for any given project in the future. All factors in the model are subject to these variances, and as these uncertainties propagate through the calculations, broad cost arranges will ultimately limit their utility.

Vessel costs are based on dayrate contracts where the construction and financial risk is held by the developer. In most early European wind farm installations, turnkey contracts transferred the risk to the contractor, leading to higher effective dayrates. In the U.S. no standard contract models have evolved and in this analysis no attempt was made to price this risk; as a result, the expected dayrates may underestimate costs.

The model does not include mobilization costs for the export cable-laying vessel. Cable-laying vessels are often loaded at the factory, and since there are no factories for marine HVAC cables in the U.S., the vessel will likely need to be loaded in Europe. Transport costs fall outside of the system boundaries and have not been included in our analysis, and therefore the mobilization of the export cable vessel is not considered. However, these costs could be significant, perhaps in the range of 2–\$10 million.

The model was developed and parameterized based on commercial (>100 MW), monopile projects using 2–3 MW turbines installed between 2000 and 2010. Applying the model to U.S. demonstration (<20 MW) and pre-commercial (20–100 MW) projects, or non-monopile projects will require re-parameterization. Future U.S. developments are likely to use 3.6–5 MW turbines, and if vessels and installation times change, or novel techniques or structures are developed outside current practice, the model will also require re-parameterization.

At present, because no projects have commenced construction in U.S. waters, the ability to calibrate the model is limited, and thus, our confidence in the ability of

Table 9.17 Alternative parameterizations impacting foundation installation time and output time per foundation

Model	Load time, L (h)	Installation time, I (h)	Movement time, M (h)	Vessel capacity, VC	Distance to port, D (nm)	Speed, S (knots)	Output, ADJUNIT (days)
Self-transport	2	36	4	8	20	10	1.9
Self-transport	3	48	6	6	50	8	2.6
Self-transport	3	60	6	4	100	6	3.5
Self-transport	3	72	6	4	50	8	3.8
Self-transport	4	72	8	4	150	6	4.3
Self-transport	4	96	8	2	150	4	6.6
Barge	NA	36	4	NA	NA	NA	1.9
Barge	NA	48	6	NA	NA	NA	2.5
Barge	NA	60	6	NA	NA	NA	3.0
Barge	NA	72	8	NA	NA	NA	3.7
Barge	NA	72	6	NA	NA	NA	3.6
Barge	NA	96	8	NA	NA	NA	4.8

Note Weather uptime W = 90%; empirical data from European project data yield a range of 1.8–5.5 days and an average of 3.6 days

the model to accurately predict installation cost is similarly limited. Prior studies have found installation costs to account for approximately 20% of total capital expenditures. The results of our model are lower than these estimates and may reflect differences in the definition of installation costs or inflation in total capital expenditures. Cost estimates are often prone to optimism. Optimism in this model may be due to underestimation of weather and geotechnical risk or work durations. Additionally if the installation vessel market is tight or fewer vessels are built or delivered later than expected, dayrates can increase and impact project costs.

Appendix A. Parameter Verification

A.1 Temporal Parameters

Empirical data on the unit time to install turbines and foundations at existing European wind farms are compared against the weather-adjusted time per unit (ADJUNIT) output from the model to confirm the parameterization. In Table 9.17 the output of a range of assumptions from Table 9.7 are used to estimate the per foundation installation time. In Table 9.18, similar assumptions from Table 9.9 are used to determine the per turbine installation time. The parameters were chosen to give a complete range of installation times and the empirically estimated times for foundation estimation range from 1.3–5.5 days with an average of 3.6 days, and for turbine installation from 1.3–9.5 days with an average of 4.0 days. In both cases, the model output is close to the mean empirical value and the range of values generated by the model matches closely with the range of values in the empirical data.

Table 9.18 Alternative parameterizations impacting turbine installation time and output time per turbine

Model	Load time, L (h)	Installation time, I (h)	Movement time, M (h)	Vessel capacity, VC	Distance to port, D (nm)	Speed, S (knots)	Model output, ADJUNIT (days)
Self-transport	2	36	4	8	20	10	2.0
Self-transport	2	48	6	6	50	8	2.8
Self-transport	4	60	6	4	50	8	3.4
Self-transport	4	72	6	4	50	8	4.0
Self-transport	4	72	6	4	100	6	4.3
Self-transport	4	72	8	4	150	6	4.6
Self-transport	6	96	8	2	150	4	7.1
Self-transport	6	120	8	2	150	4	8.3

Note Weather uptime W = 85%; empirical data from European project data yield a range of 1.3–9.5 days and an average of 4.0 days

A.2 Output Costs

Installation contracts are proprietary and are generally not publicly available, however, in three recent UK projects for turbine installation using SPIVs (Walney, Sheringham Shoal, and Greater Gabbard), cost data were made public and is used as a check on parameterization. In cases where the model output deviates from reported data, adjustments to the model parameters to account for known factors yield reasonably accurate cost. This illustrates the unique nature of each project and the need to account for individual project attributes through alternative parameterizations or adjustment factors.

Walney
Seajacks signed a $70 million turbine installation contract at the Walney windfarm [14]. Walney is composed of 102, 3.6 MW turbines and is 40 nm from port. The effective contract dayrate is $130,000 per day. When these data are input into the model (assuming expected values when not otherwise provided), the average turbine installation time is 3.2 days, the total installation time is approximately 11 months and the estimated cost is $45.9 million. The difference in cost between the model output and the contract is due to the installation time discrepancy. While both the model and the contract have a similar turbine installation time (1–4 per week in the contract and 2.2 per week in the model), the model gives a total work time of 11 months while the contract installation time is 18 months. This discrepancy is due to expected high weather downtime through the winter [15]. When the weather factor is set at 50%, the estimated cost is $74 million and the total work duration is 17.5 months, consistent with the published contract cost and duration. While there will be weather delays for U.S. projects, they may not be as severe as those experienced in Europe.

Sheringham Shoal
Master Marine signed a €78 million (approximately $101 million using 2009 exchange rates) contract for turbine installation at the Sheringham Shoal wind farm. Sheringham Shoal comprises 88, 3.6 MW turbines and is 20 nm from port. The effective contract dayrate is $374,000 per day. Using the default parameters yields an output of $37 million, however, when the dayrate is changed to reflect the contract dayrate, the model output increases to $104 million. This suggests that while the model dayrate is too low in this case, the temporal parameters are reasonable. The model dayrate is likely too low because of the size and capability of Master Marine's vessel and high competition for vessel services in the European market; a sister vessel of the one used at Sheringham Shoal received a three-year contract for work in the oil industry at a dayrate of $300,000/day. We expect that dayrates exceeding $300,000/day will be rare in the U.S. market and will only occur if vessel supply is severely constrained.

Greater Gabbard
Seajacks signed a $62 million contract to install turbines at the 504 MW Greater Gabbard wind farm. Greater Gabbard is composed of 140, 3.6 MW turbines and is 20 nm from port. The expected installation time is 14 months and the effective dayrate is $147,000/day. Using model default values yields a cost of $61.1 million and work duration of 13.9 months. The dayrate from the contract terms gives an estimated cost of $66.7 million. In this case, the model output and contract costs match closely.

References

1. Howe B (2001) Cable properties. Neptune Power. http://neptunepower.apl.washington.edu/documentation/documentation.html. Accessed 24 Oct 2010
2. de Alegría I, Martín J, Kortabarria I, Andreu J, Ere OP (2009) Transmission alternatives for offshore electrical power. Ren Sustain Eng Rev 13:1027–1038
3. Wright SD, Rogers AL, Manwell JF, Ellis A (2002) Transmission options for offshore wind farms in the United States. In: Proceedings of the American Wind Energy Association Annual Conference, 2002, pp 1–12
4. Green J, Bowen A, Fingersh LJ, Wan Y (2007) Electrical collection and transmission systems for offshore wind power. Offshore Technology Conference. 30 April– 3 May 2007, Houston, TX. 19090-MS
5. USACE (2004) Cape Wind draft environmental impact statement. U.S. Army Corps of Engineers. Concord, MA
6. Carter M (2007) North Hoyle offshore wind farm: design and build. In: Proceedings of the Institution of Civil Engineers, Energy 160: pp 21–29
7. Zaaijer MB, Van der Templ J (2004) Scour protection: a necessity or a waste of money? In: Proceedings of the 43rd IEA topical expert meeting—critical issues regarding offshore technology and deployment, pp 43–51. Skaerbaek, Denmark
8. MMS (2008) Cape Wind final environmental impact statement. Minerals Management Service. Herndon, VA. 2008-040
9. Bluewater Wind (2010) Delaware. http://www.bluewaterwind.com/delaware.htm. Accessed 24 Oct 2010
10. Musial W, Ram B (2008) Status of offshore wind energy projects, policies and programs in the United States. Presented at the 2007 European Offshore Wind Conference, Berlin. December 4–6, 2008. NREL/CP-500-42452
11. Marine Log (2010) Aker Philadelphia eyes wind turbine installation vessels. http://www.marinelog.com/DOCS/NEWSMMIX/2010feb00220.html. Accessed 24 Oct 2010
12. Schellstede H (2008) Wind power: wind farms of the northern Gulf of Mexico. Paper presented at Offshore Technology Conference, 30 April–3 May 2007. Houston, 19092-MS
13. Coastal Point Energy (2010) Introduction to Coastal Point Energy. http://www.millergroupofcompanies.com/presentations/CoastalPointEnergyPPT3.pdf. Accessed 24 Oct 2010.
14. Blackwell B (2009) DONG Energy snaps up contracts for Walney. RechargeNews.com. June 30, 2009. http://www.rechargenews.com/business_area/finance/article182126.ece?print=true Accessed 24 Oct 2010
15. DONG Energy (2010) Walney offshore wind farm. http://www.dongenergy.com/Walney/Pages/index.aspx. Accessed 24 Oct 2010

Chapter 10
Decommissioning Regulations and Workflows

10.1 Decommissioning Regulations

10.1.1 Regulatory Requirements

Energy companies that operate offshore are obligated to remove all structures, clear the site, and verify clearance upon lease termination. In federal waters, regulations for decommissioning OCS renewable energy facilities and associated structures are described in 30 CFR, Part 285, Subject I, 285.900-913. All facilities, including pipelines, cable, and other structures and obstructions must be removed when they are no longer used for operations but no later than 2 years after the termination of the lease, ROW grant, or RUE grant. Requirements for facility removal are described in 285.910:

- All facilities must be removed to a depth of 15 ft below the mudline.
- Within 60 days after a facility is removed, the site must be cleared and clearance must be verified.

The operator may request that certain facilities authorized in the lease or grant be converted into an artificial reef or otherwise toppled in place (285.909). As in the offshore oil and gas industry, all co-lessees and co-grant holders of renewable energy leases are jointly and severally responsible for meeting decommissioning obligations (285.900).

In state waters, state laws regarding decommissioning requirements are not yet defined but are expected to broadly follow OCS requirements. Timing specifications, removal depths, bonding, and clearance requirements are expected to vary depending on the state.

From the operator's point of view, decommissioning activities represent a cost to be incurred in the future, while from the government's perspective, decommissioning represents an uncertain event and financial risk if the operator becomes insolvent or cannot meet its financial obligations under the lease. For this reason,

M. J. Kaiser and B. F. Snyder, *Offshore Wind Energy Cost Modeling*,
Green Energy and Technology, DOI: 10.1007/978-1-4471-2488-7_10,

state and federal governments require companies to post a decommissioning bond at the time of construction to ensure that adequate funds exist to remove infrastructure in the future. The level of the bond is usually set at the expected cost to decommission the facilities and may be inflation-adjusted.

10.1.2 Decommissioning Bonds

Before the BOEMRE issues a commercial lease or approves an assignment of a commercial lease, the lessee must provide a $100,000 financial assurance which is to be adjusted using the Consumer Price Index-All Urban Consumers (CPI-U) over a 5-year period. For a limited lease, ROW grant or RUE grant, the lessee must provide a $300,000 CPI-U adjusted minimum lease or grant-specific financial assurance.

Before the BOEMRE approves a site assessment plan (SAP), the lessee must provide a supplemental bond or other financial assurance in an amount determined by the BOEMRE to ensure that all lease obligations are covered, including: the projected amount of rent[1] and other expected payments due to the Government over the next 12 months; any past due rent and other payments; and the estimated cost of decommissioning met-towers and related facilities as described in the SAP.

Before the BOEMRE approves a construction and operations plan (COP), the lessee must provide a supplemental bond or other financial assurance in an amount determined by the BOEMRE based on the complexity, number, and location of all facilities involved in planned activities and commercial operation. The supplemental financial assurance is in addition to the lease-specific bond and, if applicable, the SAP bond.

Before the BOEMRE allows a lessee to install facilities under the COP, the lessee must provide a decommissioning bond or other financial assurance in an amount determined by the BOEMRE based on the expected decommissioning cost. The decommissioning bond is expected to reflect the actual cost under the current technology and prices to remove all the renewable energy facilities. The BOEMRE must also approve the coverage schedule.

10.1.3 Financial Instruments

To provide evidence of financial responsibility the BOEMRE requires operators to provide insurance, a bond, a lease-specific abandonment account, U.S. Treasury

[1] Rent for OCS limited leases have been set at $3/acre. A fee of $70 for each nautical mile that a ROW crosses is also imposed, along with an additional $5/acre easement for use of the affected area.

notes, or obtain a qualified guarantor. Surety bonds have been the preferred method in the oil and gas sector to satisfy an operator's obligations and are likely to play a similar role in the offshore wind sector.

A surety bond is a guarantee that the surety will perform a duty in the event the lessee is unable to do so. For offshore wind leases, the duty is imposed by law to remove all infrastructure created by development and clear the site to ensure the performance of regulatory requirements. Surety bonds are agreements between three parties. These parties are the operator (principal) who is obligated to conduct decommissioning activities in accordance with their lease agreement, the government who acts as an agent of the landowner and is required to ensure successful operations (obligee), and an insurance company (surety) which ensures that money exists to complete decommissioning activities, regardless of the financial capacity of the principal. The surety is only responsible for the insured amount, thus the total costs of decommissioning may not be covered if due to price inflation, low capacity, unexpected circumstances (e.g., hurricane destruction), or a combination of such events, the cost of decommissioning exceeds the value of the surety.

10.2 Stages of Offshore Wind Decommissioning

Decommissioning activities are driven by economics, technology, and strategic requirements, and are governed by federal regulation. Decisions about how an offshore wind farm is decommissioned involve issues of environmental protection, safety, cost, and strategic opportunity, and the options available to developers depends upon regulatory approval and technical feasibility [1]. Deconstruction processes occur in well-defined stages in both onshore and offshore operations.

10.2.1 Project Management and Engineering

The engineering planning phase of decommissioning consists of a review of all contractual obligations and requirements from lease, operating, production, sales, or regulatory agreements. A plan is developed for each phase of the project, and the process of surveying the market for equipment and vessels is initiated. Engineering personnel assess the work requirements and the project management team will report on the options available, including the scope of work that needs to be performed and how best to prepare the bid. Permits are secured from the BOEMRE to remove structures and verify site clearance. All removal options require a disposal plan.

10.2.2 Turbine Removal

The first stage in decommissioning begins with the removal of all the wind turbines. Turbine removal is expected to follow the same process as installation, but reversed [2, 5]. After electrical isolation from the grid and removal of all lubricants, the blades,

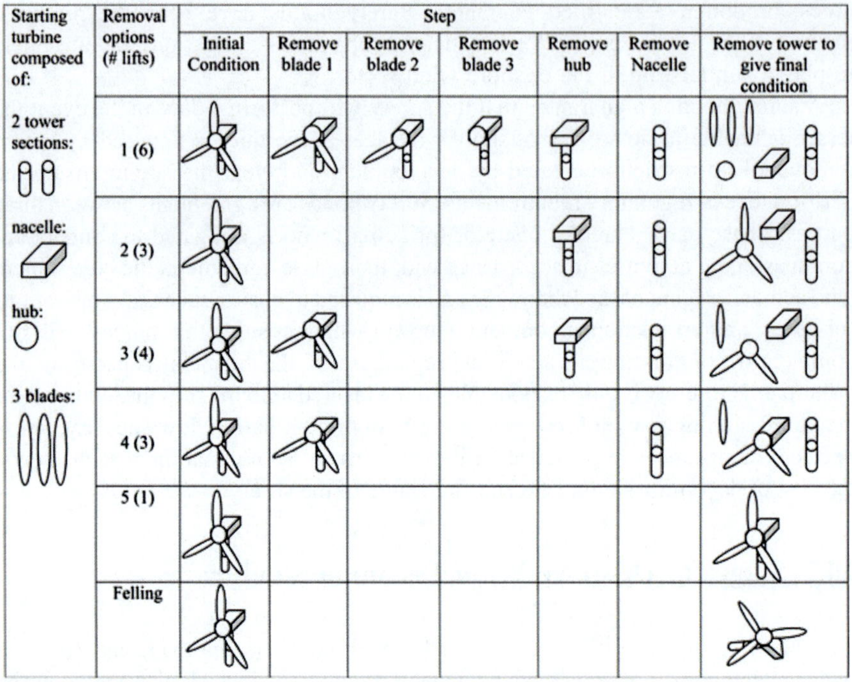

Fig. 10.1 Traditional offshore turbine decommissioning options and an alternative

hub, and nacelle will be removed, either singly or together, and the tower section will be unbolted (or cut) and removed. Removal operations occur at the end of the useful life of equipment and are not expected to be as delicate as installation, and thus, removal times are expected to be slightly less than installation times.

There are several options for turbine installation and several methods exist for removal, and given the expected high cost if vessels similar to installation are required, it is reasonable to assume that new methods will arise to ensure operations are performed efficiently and in a safe manner. The number of lifts required to remove a turbine depends on the option selected (Fig. 10.1). Vessel dayrates will vary with capacity and supply–demand conditions at the time of the operation. The same vessel spreads used in installation are likely to be required in a traditional removal because the lift weights have not changed; for a different vessel spread, the removal rates will differ.

Workflow

The general methodology is as follows:

1. Mobilize vessel and cargo barge to location.
2. Prepare turbines for cut and lift operations.
3. Remove turbine in 1–6 lifts.
4. Transport all components to an onshore site for reuse, recycling, or disposal.

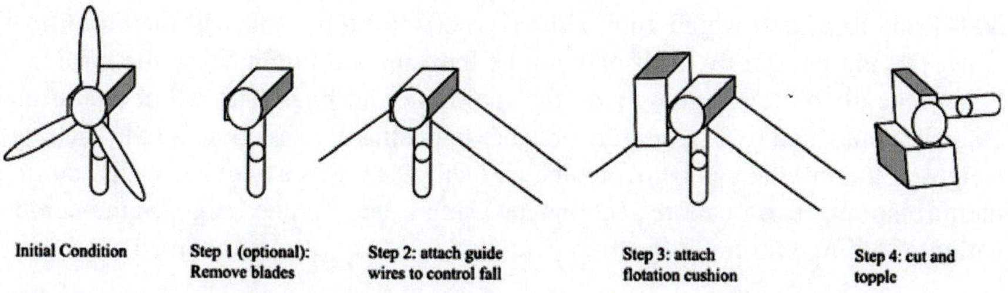

Fig. 10.2 Alternative turbine removal method

Reuse, Recycle, Disposal Options

The nacelle and hub may be disposed of in a designated landfill or they may be processed to remove steel components to sell as scrap. Blades and cable will be disposed of in a landfill. Towers will be disassembled and sold as scrap steel.

Vessel Requirements

Based on the size, weight, and height of the turbines, removal operations can be performed with vessels such as liftboats, jackup barges, or self-propelled installation vessels (SPIVs) having lift capabilities of 100–200 or more tons. In addition, cargo barges and anchor handling tugs will be required. The cost of cargo barges to transport the turbine components depends on the barge size, mob/demob time and accompanying tugs, and the amount of transported material. Support vessels are assumed to be regionally available and the distance to an onshore facility that can accept the materials will determine the transportation times.

Alternative Removal Methods

Given the high cost of turbine removal, it is reasonable to expect that creative and novel methods will be developed. One alternative method is to fell the turbine in a manner similar to cutting a tree, thereby removing costs associated with disassembly and lifting with heavy-lift vessels (Fig. 10.2). All fluids and hazardous material in the nacelle would first be removed, and the turbine tower will be cut and allowed to fall in a controlled manner into the ocean similar to methods developed for reefing-in-place oil and gas structures. The turbine would then be lifted onto a barge. The main problems to overcome include:

- Safety; the area must be clear of all personnel, and marine life must be monitored for mammals and sea turtles.
- Structural integrity; the turbine component must not hit the water with enough force to break the component.
- Flotation; the component must be kept from sinking or easily retrievable after sinking.
- Weight; a fully assembled turbine may weigh 400 tons and would require a large crane and vessel to lift onto a barge.

Ensuring flotation of the turbine can be accomplished by making the tower and nacelle watertight. A watertight, submerged turbine tower and nacelle will displace

500–1,000 tons, and weigh approximately 300–500 tons, and will therefore float. However, the nacelle by itself may not be buoyant and might cause the turbine to float vertically which might require the addition of air bags. The fall of the turbine could be controlled by one or more winches opposing the direction of fall. The large weight of the turbine can be overcome by using the buoyancy of the water, cutting the turbine into two or more sections, or using a large crane barge. Some combination of felling and traditional removal methods may also be adopted.

10.2.3 Foundation and Transition Piece Removal

After the wind turbine is removed, the foundation and transition piece will be removed. It is likely that a vessel different from the turbine removal vessel will perform foundation removals.

Cutting Method
Foundations are required to be cut 15 ft below the mudline. Mechanical casing cutters, abrasive water jets, diamond wire, or explosives can be used to make the cut at the designated elevation; selection depends upon technical feasibility, environmental conditions, regulatory options, and company preferences. Non-explosive options will likely be preferred because of the size of the monopile and the limits on explosive charges. Cutting may be performed internal or external to the monopile. Figure 10.3 depicts the processes required for an external cut; Fig. 10.4 depicts the process for an internal cut. For an internal cut, mud is jetted and pumped from the inside of the monopile to the designated below mudline depth; for an external cut, mud will be dredged around the outside of the monopile so that cutting equipment can gain access. The existence of scour protection would add cost and difficulty to an external cut, and for this reason internal cuts are considered more likely.

Lifting
If the transition piece is grouted onto the monopole, both elements will be lifted together; if the transition piece is attached in another manner, it is conceivable that separate lifts will be performed, but the only economic advantage to performing two lifts is a reduced maximum lift. A reduced maximum lift may translate into a cheaper vessel dayrate, but the incremental benefit is unlikely to account for the additional time required for the operation. A portion of the foundation will remain below the seabed; the transition piece will be removed in its entirety. The weight of the lift is based on the weight of the transition piece, grout, and the cut foundation. The foundation and transition piece will be placed on the removal vessel or cargo barge and returned to shore or to an alternative disposal site.

Workflow
The general methodology for an internal cut is as follows:

1. Gain access to the inside of the pile and set up jetting and pump equipment.
2. Pump mud from inside of monopole.

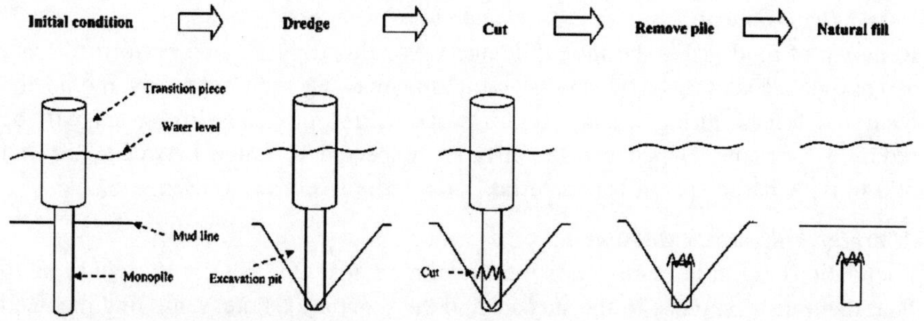

Fig. 10.3 Foundation removal—external cut

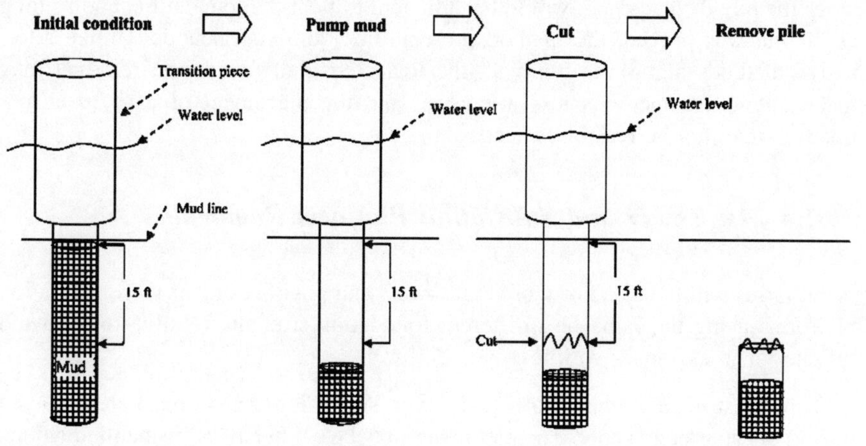

Fig. 10.4 Foundation removal—internal cut

3. Cut the monopile 15 ft below the mudline.
4. Cut turbine cable at mudline using diver support.
5. Remove transition piece and monopile foundation using suitable heavy-lift vessel.
6. Transport to onshore location for offloading/disposal or offshore reef site.
7. Remove internal equipment and disassemble onshore.

Reuse, Recycle, Disposal Options

Internal equipment is removed and disassembled and recycled or disposed in an approved landfill. Ancillary material (handrails, boat landing, ladders, etc.) will be disassembled and recycled as scrap steel. Monopiles and transition pieces will be cut into smaller pieces for sale to mini-mills if economic, disposed of in an approved landfill, or transported to an offshore reef site.

Vessel Requirements
Removal of mud inside the monopile and internal cutting may be performed from a workboat. Excavation by dredge or jetting may be required to overcome the frictional forces during lift. A jackup barge, SPIV, or heavy-lift vessel will be required for removal because the lifts are expected to range between 200 and 500 tons. A barge spread for material transport is likely to be used.

Alternative Removal Methods
Foundations could be removed using novel methods. One possibility would be to float the transition piece to the surface and then tow it to shore, removing the need for a heavy-lift vessel. This would require pumping the mud out of the pile, cutting the pile, placing an internal water barrier above the cut line, and capping the top of the transition piece. The top of the transition piece would be connected to a boat or winch and the foundation could be pulled out of the ground and floated. In most cases the pile displacement would be sufficient to float the foundation, and as long as the cut was proved, the method presents no additional hazards. Unlike alternative methods of turbine removal, the lower costs of foundation removal may make alternative methods less attractive, and the placement of caps to ensure flotation may not be feasible or economic.

10.2.4 Met Tower and Substation Platform Removal

Wind farms will have one or more met towers and possibly one or more substation platforms using the same or a different foundation structure relative to the wind turbines. The expected workflow is as follows:

1. Topside structures (meteorological tower, substation transformer) are removed and transported to shore. The met tower may be cut in half or dismantled whole, while the transformer will be lifted whole.
2. For monopile foundations, the monopile will be cut 15 ft below the mudline and removed. For jackets, leg piling will be cut 15 ft below the mudline, and then the piling and jacket will be lifted and removed in a single lift.
3. Transport the structure and superstructure to an onshore or offshore location for offloading/disposal/reefing.
4. Recycle, scrap, and landfill.

10.2.5 Cable Removal

The BOEMRE regulations for power cable and pipeline decommissioning are described in 30 CFR 250.1750-250.1754. OCS pipelines may be left in place when they do not constitute a hazard to navigation, commercial fishing, or unduly interfere with other users. State regulations are similar, but some states require the removal of pipeline segments in the surf zone to a depth of −15 ft mean lower low water (MLLW).

Regulations on renewable energy power cable decommissioning are not well developed but several options are possible:

- Complete removal of all inner-array and export cable through tidelands boundary.
- Leave in place all inner-array and export cable.
- Leave in place a portion of the inner-array and export cable.

The decision to remove cable or abandon in place determines the scope of work. Power cables may traverse a range of environmental settings, and because burial conditions may change, different solutions are likely to be required. Cable removal will likely be the preferred option when they might interfere with commercial trawling or other activities, they are located in water depths less than −15 ft MLLW or onshore and not deeply buried, or they are located in areas subject to maintenance dredging [6]. In most cases, power cables will not constitute a hazard to navigation, commercial fishing, or unduly interfere with other uses, and may be allowed to remain in place.

The removal process involves divers and/or an ROV attaching the cable to a recovery winch. The cable end is retrieved via an engine to drive the cable up onto the recovery vessel. A hydraulic shear is used to section the cable for storage and transport. Inner-array cables come in relatively short segments and can be recovered in one piece; if export cable is to be recovered in one piece, a large reel would be needed, or the cable can be cut into pieces as it is recovered. Recovery is not expected to pose technical problems under typical burial conditions due to the high strength of the cable; in some cases, water jetting may be required to fluidize sediment [7]. In situ abandonment of the inner-array cable would involve cutting the cable at the base of each turbine foundation and bury to a depth of 3 ft. If it is not practical to bury the ends, a concrete cover or mattress may be employed. For export cable a similar procedure would be employed, but at the onshore transition the cable will need to be cut and removed.

10.2.6 Scour Protection

It is expected that in many cases scour protection will remain in place in order to minimize disturbance to the seabed and provide a substrate for invertebrates. If scour removal is required, it will be conducted by a mechanical dredge or crane vessel [8]. Scour protection will need to be removed if mandated by regulators or if an external cut is required.

10.2.7 Site Clearance and Verification

The last stage in decommissioning is site clearance and verification. Site clearance is the process of removing or otherwise addressing potentially adverse impacts from debris and seafloor disturbances due to offshore wind facilities. Verification is the process of ensuring that site clearance activities are complete.

The area encompassed by a commercial wind farm is large, ranging from 0.1 to 0.3 sq km per MW capacity, so that a 300 MW wind farm would be expected to cover 30–90 km^2. Over a time frame that may exceed 20 years, debris accumulates. Piecemeal salvage is not cost effective and as long as the lost material poses no risk, debris is allowed to collect over the life of the lease. Once the facilities are removed, however, regulations and lease terms require that the location be left in a state similar to its initial development.

In oil and gas operations, regulations specify clearance based on a fixed radius per structure type. In offshore wind farms, there may be up to one hundred or more wind turbines distributed throughout a given area. BOEMRE regulations do not currently specify the area encompassed by the foundation structures. One option is to define the radius of clearance based on the structure function (met tower, foundation with turbine, foundation with substation) as in oil/gas operations.[2] Another option is to define the clearance area as the entire wind farm defined by the convex hull created by all the offshore facilities plus a buffer zone on the boundary of the convex hull. A zone covering the path of the export cables may also be included in defining the area of a wind farm.

Current regulations specify that the operator has 60 days from the time an individual structure has been removed to clear the site and verify clearance. An alternative option that may minimize administrative and contractor reporting burden would be to allow clearance to be performed 60 days after the last structure has been removed. There are economies expected to be associated with performing site clearance operations at one time and when other marine vessels are not in the operating theater.

In oil and gas operations, the BOEMRE preferred verification technique is to drag a standard trawl net across 100% of the site in two directions. In some cases, alternative site verification techniques such as side scan sonar or documentation of sweep assembly results may be used. Given the large area of wind farms and the different pattern and magnitude of debris accumulation, it may be reasonable to use or develop alternative methods of site clearance to minimize the potential environmental impacts associated with trawling operations.

10.2.8 Material Disposal

There are four methods of disposal for steel and other materials associated with an offshore wind turbine: refurbish and reuse, scrap, dispose of in designated landfills, or place offshore as an artificial reef.

Opportunities for refurbishing and reusing turbine components, foundations and transition pieces, and power cable are expected to be extremely limited due to the

[2] For comparison, the radii used in offshore oil and gas regulations are 600 ft for caissons and 1,320 ft for platforms.

age of the components at the time of removal, the nature of its assembly, and the corrosion arising from operating in a marine environment.

Scrapping and recycling are only viable for steel components,[3] but not all steel components will be economic to be recycled. Cutting steel is expensive, and if the cost to cut and transport steel in saleable units is greater than its resale value, this option will only be pursued if cheaper than the landfill option.

Recycle value depends on the weight of the component, cutting cost, scrap steel price, and transportation cost. The transition piece is usually grouted onto the monopile, and so the cost to break the grout adds to the cost to cut the pieces into segments suitable for recycle. In oil and gas decommissioning, scrap often accumulates at shipyards until scrap steel prices increase to allow a profit to be made [9]. It is not clear whether the same pattern will hold in other regions where land costs may neither be as abundant nor cheap as in South Louisiana.

If a component is not refurbished or scrapped, it will need to be disposed of in a designated landfill. Blades, power cable, some of the nacelle, grout, and marine growth are unlikely to have any resale value and will need to be disposed at a landfill. The cost to dispose depends on the processing cost, transportation cost, and disposal fee. Most of the material from a wind farm is inert and can be disposed of easily; some material is hazardous, including lubricants and electronics. Some material used in offshore wind farms could be used to make artificial reefs. Singly, monopiles lack the structural complexity required for an artificial reef, however, if a number of monopiles are positioned to achieve structural complexity, then the artificial reef option may be feasible. Other than foundations, no other material would be suitable for reefing.[4]

References

1. Department of Trade and Industry (2006) Decommissioning of offshore renewable energy installations under the Energy Act of 2004. Department of Trade and Industry, London
2. Statoil (2010) Decommissioning program. Document number sc-00-NH-F15-00005
3. London Array Limited (2009) Decommissioning programme for London Array. Document number 45-02682
4. Greater Gabbard Offshore Winds Ltd (2007) Decommissioning programme. Document number 577000/403–MGT100–GGR–107
5. Thanet Offshore Wind Farm (2008) Offshore decommissioning plan
6. Culwell AS, McCarthy J (1998) Pipeline and power cable decommissioning. In: Manago F, Williamson B (eds) Proceedings: Public workshop, decommissioning and removal of oil and gas facilities offshore California. MMS OCS Study 98-0023

[3] It is also possible that copper or magnets may be scrapped or recycled, however, the proportional weight of these materials is small.

[4] Towers could be used in reef construction but would be easily scrapped.

7. Januário C, Semino S, Bell M (2007) Offshore windfarm decommissioning: a proposal for guidelines to be included in the european maritime policy. Presented at EWEC 2007: European Wind Energy Conference, Milan
8. MMS (2008) Cape Wind final environmental impact statement. Minerals Management Service. Herndon, VA. 2008-040
9. Kaiser MJ, Pulsipher AG (2010) Scrap and storage markets in the Gulf of Mexico-III. Process workflows and industry structure. Energy Sources Part B: Econ Plan Policy 5(3):250–260

Chapter 11
Structure Weight Algorithms

11.1 Weight Algorithms

11.1.1 Tubular Weight Algorithm

Most of the material in a wind farm is composed of tubular steel elements, which is the simplest component to estimate weight since the geometric shape is uniquely described by its diameter, length, and thickness. For steel density of 7.85 g per cubic cm, the formula:

$$W = 24,660(D - t)tL \tag{11.1}$$

gives the total weight W (kg) in terms of the outside diameter D (m) of the member, wall thickness t (m), and total length L (m). For tapered components and non-uniform thickness, average diameters and thicknesses can be employed for a more accurate estimate.

In Table 11.1, unit weight in kg per linear meter is described for a tubular member in terms of outer diameter (m) and thickness (mm). The most common ranges of weight for monopiles are identified in bold. In Table 11.2, monopile weight was computed via formula and compared to reported weight at selected wind farms. On average, the estimated weight is within 10% of the reported weight. Deviations are due to non-uniform taper and thicknesses, unknown/unreported thickness, and reporting error.

11.1.2 Foundations

Foundation lift weight during removal consists of the cut monopile section, the transition piece and grout, and marine growth. We assume marine growth is negligible or has been removed. The removed foundation is composed of a section

M. J. Kaiser and B. F. Snyder, *Offshore Wind Energy Cost Modeling*,
Green Energy and Technology, DOI: 10.1007/978-1-4471-2488-7_11,

Table 11.1 Steel weight per diameter and wall thickness (kg/m)

	Wall thickness (mm)					
Nominal size (m)	12.5	25.0	37.5	50.0	62.5	75.0
0.75	227.3	447.0	658.9	863.1	1059.6	1248.4
1.2	366.0	724.4	1075.0	1418.0	1753.2	2080.7
1.5	458.5	909.3	1352.4	1787.9	2215.5	2635.5
2.5	766.8	1525.8	2277.2	3020.9	3756.8	4485.0
3	920.9	1834.1	2739.6	3637.4	4527.4	5409.8
4	1229.1	2450.6	**3664.3**	**4870.4**	**6068.7**	7259.3
4.5	1383.3	2758.8	**4126.7**	**5486.9**	**6839.3**	8184.0
5	1537.4	3067.1	**4589.1**	**6103.4**	**7609.9**	9108.8
5.5	1691.5	3375.3	**5051.4**	**6719.8**	**8380.5**	10033.5

Note The most common range of weights for monopiles are identified in bold

Table 11.2 Specifications and estimated weights of monopiles at selected wind farms

			Wall thickness		Weight	
Wind farm	Length (m)	Outer diameter (m)	Reported (cm)	Assumed (cm)	Reported (ton)	Estimated (ton)
Kentish Flats	38–44	4.3	4.5		144–184	178–206
Horns Rev	34	4	5		160	164
Horns Rev 2	30–40	3.9	4.0–8.2		150–210	169–226
North Hoyle	50	4	3.0–7.0	5	250	242
OWEZ	45	4.6	4.0–6.0	5	230	250

of "free" monopile, a section of monopile grouted to the transition piece (referred to as monopile/transition overlap), and a section of free transition piece. See Fig. 11.1. The weight of each component must be determined because the processing and disposal costs of grouted tubular components are different from the costs of free components.

11.1.2.1 Monopile

Monopile weight is frequently reported in public documents, but in decommissioning operations it is the weight of monopile removed that is the important factor. Federal regulations require monopiles to be cut 4.6 m (15 ft) below the mudline and the total weight of the monopile removed is composed of the monopile below the mudline, the monopile from the mudline to the water level, and the monopile above the water line.

The portion of monopile grouted to the transition piece usually ranges from 1.3 to 1.6 times the pile diameter [1, 2]. Given the length of monopile/transition overlap and the total length of the monopile, the length of free monopile is the total length minus the length of monopile grouted to the transition piece. The weight of

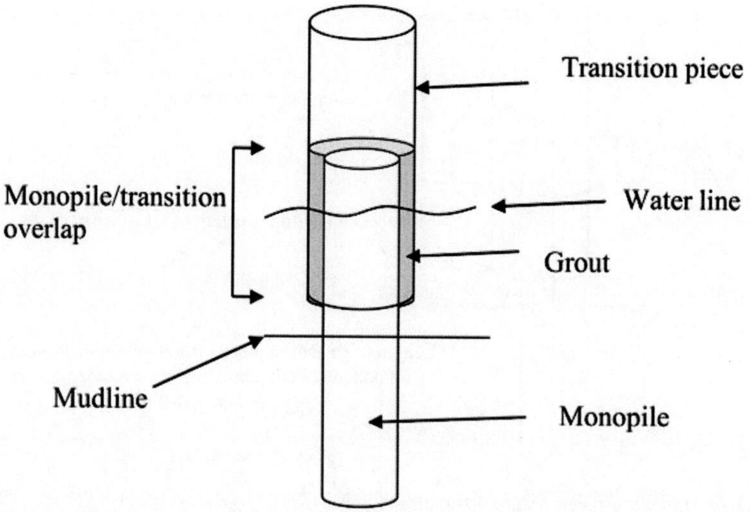

Fig. 11.1 Components of a wind turbine foundation

Table 11.3 Monopile dimensional specification estimates

Specification	Unit	Calculation
Diameter (D_M)	m	
Length of monopile above water (L_MW)	m	
Length of total monopile removed (L_TMR)	m	WD + 4.6 + L_MW
Monopile/transition overlap factor (OF_MT)		
Length of monopile/transition overlap (L_MT)	m	OF_MT * D_M
Length of free monopile (L_FM)	m	L_TMR–L_MT
Wall thickness (WT_M)	m	
Unit weight (UW_M)	ton/m	
Weight of total monopile removed (W_TMR)	ton	(WD + L_MW+4.6)*UW_M
Weight of free monopile (W_FM)	ton	UW_M * L_FM
Weight of monopile in mon/trans (W_MMT)	ton	L_MT * UW_M

free monopile and the weight of monopile grouted to the transition piece is determined by the length times the unit weight.

Table 11.3 summarizes the monopile removal calculations and assumes a 4.6 m below mudline cut, that the length of monopile grouted to the transition piece is 1.3–1.6 times the pile diameter, and a wall thickness of 35–65 mm. The length of the monopile above water is assumed to range from 1 to 5 m and the unit weight ranges from 2.7 to 7.6 ton/m.

To illustrate application of Table 11.3, consider a 4.5 m diameter monopile in 10 m water depth with a wall thickness of 50 mm. Assume the pile extends 1.5 m above the water line. The total length of the monopile removed is the water depth, plus the 4.6 m below mudline, plus the length of monopile above the waterline, or 16 m. The weight of the total monopile removed is the length times the unit weight

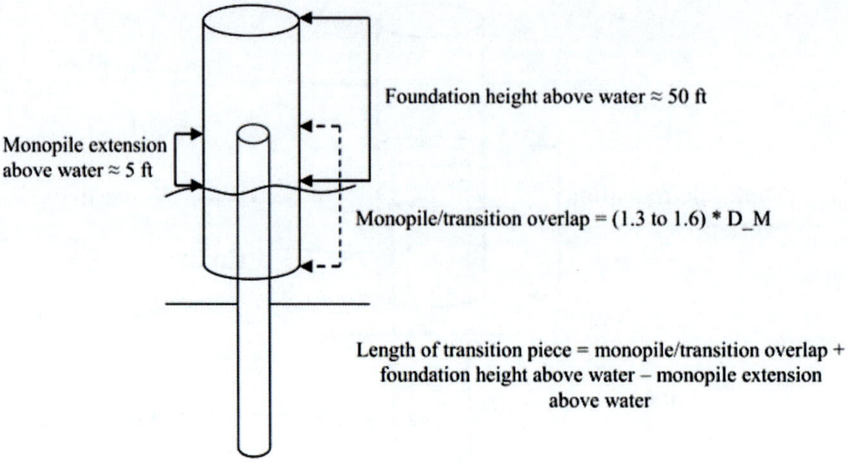

Fig. 11.2 Transition piece height estimation

(5.48 ton/m) and total weight is therefore 16 m*5.48 ton/m = 87.8 ton. The length of monopile grouted to the transition piece is 4.5 m*(1.3 to 1.6) = 5.9 to 7.2 m and the length of the free monopile is 16 m–(5.9 to 7.2 m) = 8.8 to 10.1 m. Weights are the length times the unit weight.

11.1.2.2 Transition Piece

If length, diameter and wall thickness of the transition piece is known, weight may be estimated from the tubular steel weight formula. If unknown, values may be estimated as a function of monopile diameter and height above water. Figure 11.2 illustrates the transition piece length estimation method and Table 11.4 gives the calculations, assumptions, and nomenclature.

We consider a 4.5 diameter monopile that extends 1.5 m above the water for illustration. Assume the transition piece has a wall thickness of 50 mm, a height above the waterline of 15 m, and an annulus of 50 mm. The length of the transition piece is the length of the overlap, plus the length above water, minus the length of the monopile above water or: (1.3 to 1.6)*4.5 + 15 − 1.5 m = 19.4 to 20.7 m. The diameter of the transition piece is the diameter of the monopile, plus the annulus and the wall thickness of the pile or: 4.5 m + 2*(0.05 m) + 2*(0.05 m) = 4.7 m. We estimate the unit weight as 5.8 ton/m. Multiplying unit weight by the length gives 5.8 ton/m*(19.4 to 20.7 m) = 112 to 120 ton.

Specifications of transition pieces for select wind farms and estimated weights from the Table 11.4 formula are shown in Table 11.5. Wall thickness was assumed to be 50 mm and the annulus 125 mm; if height above water was unknown, it was assumed to be 15 m. The formula weight is generally lower than the reported weight, which may be due to the presence of secondary steel or

Table 11.4 Transition piece dimensional specification estimation

Specification	Unit	Assumption	Calculation
Height above water (H_T)	m	8–20 (mean 15)	
Annulus (A_T)	m	0.05–0.12	
Length (L_T)	m		O_MT + H_T − L_MW
Diameter (D_T)	m		D_M + (A_T)*2 + (WT_T)*2
Wall thickness (WT_T)	m	0.0375–0.0625	
Unit weight (UW_T)	ton/m	2.7–7.6	
Weight (W_T)	ton		UW_T * L_T

Table 11.5 Weights of transition pieces at selected wind farms

Wind farm	Outer diameter (m)	Weight (ton)	Length above water (m)	Total length (m)	Estimated weight (ton)
Kentish Flats	4.5	90	8		92
Burbo Bank	5	225	22		194
Belwind	4.3	160		22	137
Gunfleet Sands	5	212	18		169
Lincs	5	250	6		86
OWEZ	4.3	250	13	25	156
Lynn/Inner Dowsing	5	181	21		188
Rhyl Flats	5	220	23		200
Robin Rig	4.5	160	20		177
Horns Rev 2	4.2	170			139
Baltic 1	4.4	250		27	169

concrete in the boat landings, tidal differences in the measurement of height above water, or reporting differences. On average, the formula underestimates reported weight by 33%.

11.1.2.3 Grout

Grout is used to secure the transition piece to the monopile.[1] The weight of grout is the volume of grout used multiplied by its unit weight. Figure 11.3 illustrates the grout volume estimation. The volume of grout is computed as the volume difference created by the outside diameter of the monopile and the inside diameter of the transition piece. The length of both cylinders is given by the overlap (1.3–1.6 times the diameter); the inside diameter of the transition piece is given by the outer diameter of the monopile plus twice the annulus.

1 In recent years, problems with grouting at some sites have emerged which have prompted redesign and costly intervention. This may lead to changes in grouting techniques and foundation design which could influence weight estimates.

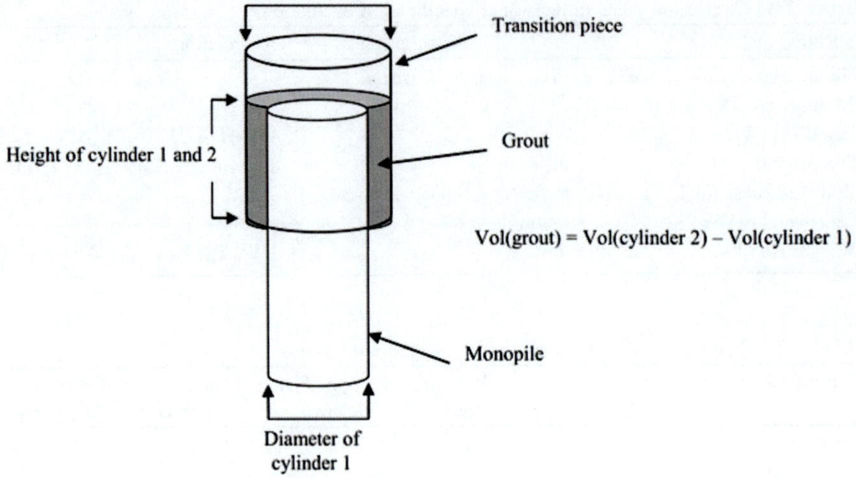

Fig. 11.3 Grout weight estimation

Table 11.6 Weight of cement grout used in turbine foundations

Pile diameter (m)	Pile overlap as a function of pile diameter	Annulus width (mm)	Grout volume (m^3)	Weight (ton)
4	1.3	50	1.7	3.7
5	1.3	50	2.6	6
4	1.6	125	4.9	11
5	1.6	125	7.8	17.5

Note The density of cement is assumed to be 2,500 kg/m^3

Table 11.6 shows estimates for grout weight given different assumptions on annulus size and pile-transition piece overlap. In all cases, the weight of grout is small relative to other weights in the system and grout is not expected to exceed 10% of the total weight removed.

11.1.3 Tower and Turbine

If tower height, diameter, and wall thickness are known or can be reasonably approximated, weight may be calculated via formula. If these variables are not known, they may be estimated from the turbine specifications shown in Table 11.7. The ranges of tower weight for a given turbine are due to differences in hub height. In Table 11.8, the tower length and weight at selected offshore farms are depicted. As the capacity of a turbine increases, weight and load increases, and along with hub height, impacts tower weight [3]. Differences in technology and design account for the variation observed.

Table 11.7 Tower weights by turbine type

Turbine	Capacity (MW)	Tower (ton)
Vestas V80	2	130–200
Vestas V90-3	3	100–150
Siemens 3.6-107	3.6	180–200
Repower 5 M	5	210–225

Table 11.8 Tower length and weight at selected offshore wind farms

Wind farm	Turbine	Tower weight (ton)	Tower height (m)
Kentish Flats	Vestas V90-3	108	62
Horns Rev	Vestas V80	160	
Gunfleet Sands	Siemens 3.6-107	193	60
Alpha Ventus	Repower 5 M	210	
Burbo Bank	Siemens 3.6-107	180	65
Arklow Bank	GE 3.6	160	70
Beatrice	Repower 5 M	225	59

Table 11.9 Weight of turbine components in commonly used offshore wind turbines

Turbine	Capacity (MW)	Diameter (m)	Rotor (ton)	Nacelle (ton)	Total (ton)
Siemens 2.3-93	2.3	93	60	82	142
Nordex N90	2.5	90	55	91	146
Vestas V90-3	3	90	42	70	112
Siemens 3.0-101	3	101	40	73	113
Siemens 3.6-107*	3.6	107	95	125	220
Repower 5 M	5	126	120	300	420
Multibrid M5000	5	116	110	199	309
Repower 6 M	6	126	135	325	460

Note * Cape Wind proposed wind turbines
Source Turbine specification sheets; industry press

Turbine rotor and nacelle weight vary based on capacity, blade length, drive type, and manufacturer. The most popular European offshore turbines are given in Table 11.9. In recent years, major turbine suppliers have developed direct drive turbines which eliminate the use of a gearbox in the turbine. This results in significant weight savings in the turbine and may have impacts on the weights of other farm components.

Much of the weight of the turbine is composed of steel, and if the turbine is disassembled, these materials will be sold or scrapped. Processing costs will determine material disposition. Table 11.10 shows general material composition by weight. Approximately 70% of the weight of the turbine is composed of steel, some of which may be recycled, depending on processing costs and scrap prices. Steel in the hub, gearbox, and frame is relatively accessible.

Table 11.10 Material usage in large (4 MW) turbines

Component	Steel (%)	Copper (%)	Other (%)	Proportion of turbine weight* (%)
Blades	2		98	23
Gearbox	96	2	2	30
Generator	93	4	3	8
Frame	85	3	12	20
Hub	100			18
Total	72	1.5	26	100

Note * Turbine weight includes the blades, nacelle and hub
Source NREL 2008

11.1.4 Cable

Total cable weight is a function of the weight per meter times the cable length in meters. The weight of cable varies by size and capacity and inner-array cable weight may vary within a wind farm because as more turbines are connected in series, the size of the power cable must increase to handle the load. Table 11.11 describes export and inner-array cables used at selected wind farms according to high- and medium-voltage classes. Export cables generally weigh 50–100 kg/m and inner-array cables weigh 20–40 kg/m [4, 5]. Average medium-voltage cable weighs 30.5 kg/m; average high-voltage cable weighs 75.2 kg/m.

11.1.5 Substation

The wind farm substation will be supported on a monopile or jacket foundation. Jacket weight depends on framing configuration, the degree of batter, pile requirements, and topsides load. Design philosophy and soil conditions also play a role in determining the amount of steel used in construction. Jacket weight for select wind farms in different offshore regions are summarized in Table 11.12 in terms of water depth and topsides load. Topsides for offshore wind substations range in weight from 500 to 2000 tons; the BorWin1 platform is an HVDC platform and is especially heavy.

The data in Table 11.12 yields the functional relationship:

$$\text{Weight} = 12.8\text{WD}^{0.19}\text{Load}^{0.48} \tag{11.2}$$

where Weight is in metric tons, water depth (WD) is given in meters, and topside Load is in metric tons. The equation predicts about half of the variation in jacket weight. If additional data is available, subcategories can be defined and the model relation refined.

Table 11.11 Medium and high-voltage cable weights

Windfarm	Voltage (kV)	Weight (kg/m)
North Hoyle	33	37
Kentish Flats	33	32
Walney	33	34
Cape Wind	34.5	22
Cape Wind	34.5	31
Cape Wind	34.5	41
Lillgrund	36	16
Lillgrund	36	19
Lillgrund	36	22
Lillgrund	36	56
OWEZ	36	19
OWEZ	36	33
Burbo	36	35
Cape Wind	115	83
Nysted	132	72
Sheringham Shoal	132	77
Princess Amalia	150	69
Horns Rev	150	70
Belwind	150	80

Table 11.12 Jacket weights at selected offshore wind farms

Wind farm	Weight (ton)	Water depth (m)	Topside load (ton)
Walney	990	30	1,030
Alpha Ventus	750	30	700
Horns Rev 2	800	13	1,230
Generic UK	1,200	19	1,600
Generic UK	1,400	24	1,800
Lincs	950	10	2,250
Thanet	695	21	1,200
BorWin1	1,700	40	3,200
Greater Gabbard	767	30	2,069

11.2 Example

The weight relations were used to estimate Cape Wind component weights. From the Environmental Impact Statement [6], we assume the mean water depth at Cape Wind is 9.5 m, pile diameters are 5.2 m and extends 6 m above the waterline, and the transition piece extends 10 m above the water. Given these assumptions, the weights in Table 11.13 were calculated.

The total length of monopile removed is the length below the mudline (4.6 m) plus the water depth (9.5 m) plus the extension above the waterline (6 m). Approximately 6.8–8.3 m of overlap between the transition piece and monopole is

Table 11.13 Cape Wind farm components and disposal weights

Component	Length (m)	Weight (kg/m)	Weight (ton)	Total farm weight (ton)	Percentage (%)
Total monopile	20.1	6,300	127	16,510	17.9
Free monopile	11.8–13.3	6,300	74–84	9,620–10,920	
Monopile/transition	6.8–8.3	12,700	86–105	11,180–13,650	
Total transition	10.8–12.3	6,400	69–79	8,970–10,270	10.4
Grout			13.8–16.8	1,794–2,184	2.2
Tower	68.5		190	24,700	26.8
Turbine			220	28,600	31.1
Export cable	80,400	83		6,030	6.6
Inner-array cable	127,000	30		3,810	4.1
Substation jacket			800	800	0.9

required for a 5.2 m diameter pile, leaving 11.8–13.3 m of free monopile. The transition piece length is the monopile-transition overlap (6.8–8.3 m), plus the length above water (10 m) minus the monopile length above water (6 m) or 10.8–12.3 m. Assuming a wall thickness of 50 mm, the unit weights are estimated as 6,300 kg/m for monopiles and 6,400 kg/m for transition pieces. Assuming an annulus of 50 mm, approximately 14–17 ton of grout is required per foundation.

Tower and turbine weights are estimated from the data in Tables 11.7, 11.8 and 11.9. Cape Wind has reported that they plan to use Siemens 3.6 MW turbines, which, based on previous installations, will weigh approximately 220 tons and be supported by 190 ton towers. Cape Wind has also made information on the weights of export and inner-array cables available [5]. The Cape Wind substation is to be placed in shallow water (8.5 m); assuming a topside load of 1,500 tons, the jacket is expected to weigh approximately 800 tons.

The vast majority of the weight of the decommissioned components at Cape Wind will consist of steel in the turbine towers, turbines, and monopiles and transition pieces. Turbines will be the heaviest component and will account for approximately 30% of the material decommissioned. The towers and foundations will each account for 25–30% of the disposal weight. If removed, export and inner-array cable will account for approximately 10% of the weight of removed components.

References

1. Moller A (2008) Efficient offshore wind turbine foundations. Power Expo, Zaragoza, Spain, 24–25 Sept 2008. Available at: http://www.aeeolica.es/doc/powerexpo2008/06-Anders-Moller-DENSIT.pdf
2. Schaumann P, Wilke F, Lochte-Holtgreven S (2008) Nonlinear structural dynamics of offshore wind energy converters with grouted transition piece. European Wind Energy Conference. Brussels. March 31–April 3

3. Thresher RW, Dodge DM (1998) Trends in the evolution of wind turbine generator configurations and systems. Wind Energy 1:75–85
4. Wright SD, Rogers AL, Manwell JF, Ellis A (2002) Transmission options for offshore wind farms in the United States. In: Proceedings of the American Wind Energy Association Annual Conference, pp 1–12
5. Pirelli ABB (2004) 115 kV Solid dielectric submarine cable. Available at http://www.boemre.gov/offshore/PDFs/CWFiles/120.pdf. Accessed 3 Jan 2012.
6. MMS (2008) Cape Wind final environmental impact statement. Minerals Management Service, Herndon, VA. 2008-040

Chapter 12
Decommissioning Cost Estimation

12.1 Removal Options

12.1.1 Model Structure

Decommissioning stages are assumed to be performed independently using separate vessels. The removal costs for each component are calculated using a specific model and each component may have removal, processing, disposal costs, and scrap revenues. Removal costs are a function of the work duration per unit, the number of units, and the dayrates of the vessel(s) conducting the work. Processing and disposal costs and scrap revenue are calculated separately for each component on a price per ton basis. The total costs of decommissioning are the sum of the costs of all removal, processing, disposal, and scrap modules.

12.1.2 Turbines

Two traditional options and one unconventional approach for turbine removal are modeled. After removal, turbine components may be carried back to shore by the removal vessel (called "self-transport"), or components may be transferred to a barge and barged back to shore (called "barge"). The choice depends on the costs of barge spreads, removal vessel costs and capacities, and the expected duration of activity. An alternative method based on a "cut-and-fall" approach (called "felling") is also described.

M. J. Kaiser and B. F. Snyder, *Offshore Wind Energy Cost Modeling*, Green Energy and Technology, DOI: 10.1007/978-1-4471-2488-7_12,

12.1.3 Foundations

Two vessel options exist for the internal-cut method based on the vessel that supports cutting operations. In "single-vessel" operations, a SPIV or jackup supports cutting operations and is utilized to lift the foundation and place it on a barge. In "OSV support" multiple vessels are used. An OSV is used to support cutting operations and a lift vessel arrives on site only after the foundation is cut; this reduces the total amount of time the larger removal vessel is required. The OSV support option may be less expensive, but would rely on the monopile to be held in place by a short section of buried pile until the lift vessel arrived; this may or may not be technically feasible.[1] In OSV support, it is assumed that the lift vessel will not incur any downtime while waiting for piles to be cut; in practice, this may require two or more OSVs to be employed simultaneously.

12.1.4 Cable

Vessel requirements for cable removal are less severe than those for cable installation[2] and it is assumed that low cost vessels, specifically tugs, barges, mechanical dredges, and OSVs, will be utilized. All cable is assumed to be removed, although it is possible that a portion of the cable will be allowed to remain in place after environmental review. In removal, the cable is expected to be cut in several small sections. In general, the ability to remove cable without dredging or jetting depends on the weight and length of the cable, the sediment type, and the capacity of the winch. Vessels with large turntables and ROVs are not required in removal operations.

12.1.5 Substation and Met Tower

Substation decommissioning consists of a topside removal plus jacket removal. A heavy-lift vessel will be required to lift the topsides from the jacket and place it on a barge. The same vessel will perform the jacket lift after the foundation piling has been cut. If the foundation piling is grouted into the jacket legs, the foundation and

[1] In water depths less than 15 ft, the foundation above the mudline would be supported by an approximately equal amount of foundation below the mudline, which would likely be adequate to prevent toppling in normal weather conditions. In water depths that exceed 30–50 ft, this approach might not be feasible.

[2] Cable laying is more difficult than cable removal due to the vessel requirements and the need to lay cables in a single continuous piece and to bury the cable.

piling will be lifted together. A met tower will undergo a similar process but with a smaller, less expensive vessel.

12.2 Turbine Removal Model

12.2.1 Input

The user specifies the physical parameters of the wind farm including the distance to port, the farm capacity, and the turbine capacity. The user is also required to specify the removal vessel type (jackup, liftboat, or SPIV) and the logistical system (self-transport or barge) employed. The model provides dayrates, vessel capacities, spread requirements, and activity durations. Historic dayrates and current technologies are assumed. Activity durations are hypothesized relative to empirically derived installation times.

12.2.2 Self-Transport Model

In self-transport, the main vessel will remove and store turbine components until its capacity is reached, at which time it will return to port and offload components. Removal rates are reported on a per-trip basis. The total time per trip is the sum of the travel time, loading time, removal time, and intra-field movement time.

The total travel time (TRAVEL, h) is determined by the speed of the vessel (S, kn) and the distance to port (D, nm):

$$\text{TRAVEL} = 2\left(\frac{D}{S}\right) \tag{12.1}$$

Total per-trip removal time (REMOVE, h), total off-loading time (LOAD, h), and total per-trip intra-field movement time (MOVE, h) are a function of vessel capacity (VC, number turbines) and removal (R, h), off-loading (L, h), and intra-field movement (M, h) times per turbine:

$$\text{REMOVE} = \text{VC} \times R \tag{12.2}$$

$$\text{LOAD} = \text{VC} \times L \tag{12.3}$$

$$\text{MOVE} = \text{VC} \times M \tag{12.4}$$

The total time per trip (TRIP, h) is:

$$\text{TRIP} = \text{TRAVEL} + \text{LOAD} + \text{REMOVE} + \text{MOVE} \tag{12.5}$$

A weather-adjusted time per trip (ADJTRIP, h) is given by:

$$\text{ADJTRIP} = \text{TRIP} \times \left(\frac{1}{W}\right) \tag{12.6}$$

where the factor W accounts for the proportion of time vessels are unable to operate. $W = 1$ indicates that there is no weather delay; $W = 0.5$ indicates that 50% of the time vessels are unable to operate.

The number of trips (NUMTRIP) required is determined from the total number of turbines (NUMTURB) and the vessel capacity:

$$\text{NUMTRIP} = \frac{\text{NUMTURB}}{\text{VC}} \tag{12.7}$$

Removal time (REMTIME, h) is determined from:

$$\text{REMTIME} = \text{ADJTRIP} \times \text{NUMTRIP} \tag{12.8}$$

The total daily cost (TDC, \$/day) is the sum of the vessel dayrate (VDR, \$/day) and the required spread dayrate (SDR, \$/day):

$$\text{TDC} = \text{SDR} + \text{VDR} \tag{12.9}$$

Total project cost (COST, \$) is removal time normalized to days and multiplied by TDC:

$$\text{COST} = \frac{\text{REMTIME}}{24} \times \text{TDC} \tag{12.10}$$

12.2.3 Barge Model

In the barge approach, a cargo barge is assumed to always be available to accept turbine components from the removal vessel so that there is no downtime due to logistical constraints. Since the main removal vessel does not travel to and from port, travel time (TRAVEL) and loading time (LOAD) are not required and the model calculates removal time per unit.

Total turbine removal time per turbine (TURB, h) is the sum of the time to remove a turbine (R) and move to another field location (M):

$$\text{TURB} = R + M \tag{12.11}$$

The weather-adjusted time per turbine is

$$\text{ADJTURB} = \text{UNIT} \times \left(\frac{1}{W}\right) \tag{12.12}$$

and the total installation time is the product of the weather-adjusted time per turbine and the number of turbines installed:

$$\text{REMTIME} = \text{ADJTURB} \times \text{NUMTURB} \tag{12.13}$$

As before, the total cost is determined as time multiplied by the daily cost:

$$\text{COST} = \frac{\text{REMTIME}}{24} \times \text{TDC} \tag{12.14}$$

12.2.4 Unconventional Model

Alternative methods to remove turbines will be investigated by project engineers to reduce decommissioning expenditures. One potential method is to fell the turbine in a manner similar to cutting a tree, eliminating costs associated with specialized vessels and potentially reducing the time on site to perform the operation.

After fluids and hazardous material in the nacelle are removed, the turbine tower would be cut and allowed to fall into the ocean in either a controlled or uncontrolled manner. The turbine would then be lifted or floated onto a barge and brought to shore. Significant engineering work would be required to ensure the safety and efficiency of the operation.

The cost of felling is the product of the vessel dayrates and the operational time. The vessels required depend on the method of removal; at a minimum, one OSV would be required to support cutting and preparatory operations, one barge spread would be required to transport the cut turbine to port,[3] and one additional vessel would be required to perform utility work. At a maximum, these requirements may be doubled. Assuming vessel costs as before ($7,500/day for OSVs, $10,000/day for tugs, $3,500/day for crew/utility boats and $1,750/day for barges, plus a $1,000 crane cost) yields a total dayrate between $24,750 and $48,500 per day.

The duration of felling operations are uncertain, and given safety concerns, significant preparatory time may be required to remove sensitive components and affix guide lines or flotation systems. Cutting and lifting operations are expected to proceed rapidly. In total, approximately 2–4 days are expected to prepare the turbine for cutting and 1–2 days may be required to cut and recover the turbine. The total removal cost per turbine is estimated to range between $74,250 and $292,000 with an expected cost of $165,000.

[3] The barge may either be equipped with a crane or may be semisubmersible; alternatively the OSV could be equipped with a mobile crane.

Table 12.1 Vessel costs and capacities for turbine removal

Vessel type	Speed (kn)	Turbine capacity	Dayrate range ($/day)	Exp. dayrate ($/day)
Liftboat	4–6	1–2	12,500–75,000	35,400
JU barge	4–8	2–6	25,000–150,000	64,200
SPIV	8–12	6–8	60,000–300,000	134,300

Table 12.2 Spread dayrate costs for turbine removal by vessel category and transport system

Vessel type	Transport system	Number of vessels			Total dayrate ($/day)		
		Tugs	Barges	Crewboats	Min	Max	Expected
SPIV	Barge	2–3	2–3	2–4	30,500	49,250	39,875
SPIV	Self-transport	0	0	1-3	3,500	10,500	7,000
Jackup	Barge	3–4	2–3	2–4	40,500	59,250	49,875
Jackup	Self-transport	1–2	0	1–3	13,500	30,500	22,000
Liftboat	Barge	2–3	2–3	2–4	30,500	49,250	39,875
Liftboat	Self-transport	0-1	0	1–3	3,500	20,500	12,000

Note Dayrates of tugs, barges, and crewboats are assumed to be $10,000, $1,750, and $3,500 per day, respectively

Table 12.3 Turbine removal time by turbine capacity and vessel type

Turbine capacity (MW)	Vessel type	Removal time, *R*(h)		
		Min	Max	Expected
2.5–3	LB	65	86	76
	JU	43	65	54
	SPIV	32	43	38
3–4	LB	86	108	97
	JU	54	86	65
	SPIV	43	54	49
4–5	LB	NA	NA	NA
	JU	65	108	86
	SPIV	54	86	65

12.2.5 Parameterization

Vessel cost and capacity is input according to Table 12.1. The combination of vessel type and logistical system determines the spread requirements and costs according to Table 12.2.

Table 12.3 shows the removal times for turbines of varying capacities. Smaller turbines have shorter removal times due to the ability of the removal vessel to complete the job in fewer lifts. Removal time is estimated from the installation time multiplied by a discount factor (DF) which is assumed to vary from 0.75 to

Table 12.4 Parameterization range for factors influencing turbine removal

Model	Offload time, L (h)	Movement time, M (h)	Weather uptime, W (%)
Self-transport	2–4 (3)	4–8 (6)	75–90 (85)
Barge	NA	4–8 (6)	75–90 (85)

Note Expected value denoted in parenthesis

Table 12.5 Turbine removal cost estimates at the Cape Wind farm

Vessel type	Logistical system	Cost (million $)
SPIV	Self-transport	51.5
SPIV	Barge	61.0
Jackup	Self-transport	42.1
Jackup	Barge	51.6
Liftboat	Self-transport	36.5
Liftboat	Barge	49.4
Average	–	48.7
Alternative method	–	22.1

0.99 with an expected value of 0.90 [1]. The discount factor accounts for the less delicate nature of the removal operation. Current vessel capabilities preclude the use of liftboats in the removal of 4–5 MW turbines. Table 12.4 shows the model input for weather factor, movement time, and offloading time. Movement time and weather factor are assumed to be identical to installation operations, but offloading time during removal is expected to proceed slightly quicker than loading time during installation.

12.2.6 Example

Cape Wind is to be composed of 130, 3.6 MW turbines. A port near Quonset, Rhode Island, approximately 60 nm from the wind farm, will reportedly serve as the staging and delivery area. Barge and self-transport models are compared across all vessel types. All dayrates and work durations are set at their expected levels.[4]

Table 12.5 shows the results. Turbine removal costs at Cape Wind are expected to range between $35 and $60 million, depending on vessel and logistical system. These costs are slightly less (approximately 10%) than expected installation costs. Liftboats are less expensive than jackups or SPIVs and the self-transport system is less expensive than the barge system. The expected turbine removal costs using the felling method are $22.1 million.

4 If the minimum and maximum parameterizations from Tables 12.1, 12.2, 12.3 and 12.4 are used the range in costs increases dramatically and becomes uninformative.

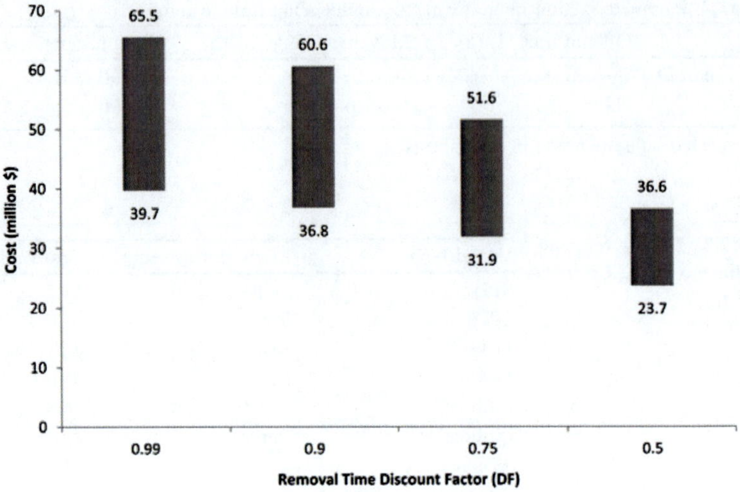

Fig. 12.1 Turbine removal cost at Cape Wind by assumed discount factor

Figure 12.1 shows the range of turbine removal costs for various discount factors. Only expected parameters are input and the range (the height of the bars) is generated by differences in vessel type and logistical method. While the discount factor is important, the vessel type and logistical method are also important and lead to similar levels of uncertainty.

12.3 Foundation Removal

12.3.1 Input

The user specifies the farm and turbine capacity, the vessel type and the logistical method (OSV support or single-vessel). Additionally, the user may provide input on pile diameter, or a default value provided by the model may be utilized.

12.3.2 Single-Vessel

The total time required in single-vessel removal is calculated on a per-foundation basis. Time to remove a foundation is composed of time to stabilize on site (JACKUP, h), time to pump mud from the foundation (PUMP, h), time to cut the foundation (CUT, h/m) for pile diameter (D, m), time to lift the foundation and place on a barge (LIFT, h), and time to jack the vessel down and move to the next foundation (MOVE, h). The sum of these times is the total time per foundation (TPF, h):

$$\text{TPF} = \text{JACKUP} + \text{PUMP} + (\text{CUT} * D) + \text{LIFT} + \text{MOVE} \tag{12.15}$$

The total cost of foundation removal is the product of the total time per foundation, the number of foundations (NUMFOUND), and the total daily cost (TDC, $/day). The number of foundations is determined from the farm capacity divided by the turbine capacity, and the total daily cost is the sum of the vessel and spread dayrates (VDR and SDR):

$$\text{COST} = (\text{VDR} + \text{SDR}) * \left(\frac{\text{FC}}{\text{TC}}\right) * \text{TPF} \tag{12.16}$$

The model does not contain a time discount for vessel type; that is, cutting time (or stabilizing time, lifting time, etc.) does not differ by vessel class and therefore the vessel with the lowest total dayrate (including spread) capable of completing the work will always be the cheapest alternative.

12.3.3 OSV Support

For OSV support, the OSV cost and the main removal vessel costs are summed. The main removal vessel daily cost is the sum of the vessel and spread dayrates. The time per foundation for the removal vessel (TPFr, h) is composed of time to stabilize at the site, time to lift the foundation and place on a barge, and time to jack the vessel down and move to the next foundation:

$$\text{TPFr} = \text{JACKUP} + \text{LIFT} + \text{MOVE} \tag{12.17}$$

The time required to cut the monopile (TPFo, h) is composed of time to jet and pump out mud and cut the pile, plus time to stabilize at the foundation (STABILIZE, h) and move to the next foundation:

$$\text{TPFo} = \text{STABILIZE} + \text{PUMP} + \text{CUT} + \text{MOVE} \tag{12.18}$$

The daily cost of the OSV is the dayrate (ODR, $/day). The time for an OSV to stabilize on site and move to the next site is assumed to be negligible. The total cost for the OSV and lift vessel is:

$$\text{COST} = \left[\text{TPFR}*\frac{\text{FC}}{\text{TC}}*(\text{VDR} + \text{SDR})\right] + \left[\text{TPFo}*\frac{\text{FC}}{\text{TC}}*\text{ODR}\right] \tag{12.19}$$

Table 12.6 Time per foundation for alternative foundation removal methods

Vessel	Task	OSV support time (h)			Single vessel time (h)		
		Expected	Min	Max	Expected	Min	Max
OSV	Stabilize	0.5	0.25	2	–	–	–
	Pump	4	2	6	–	–	–
	Cut	16 h/m	10 h/m	24 h/m	–	–	–
	Move	0.5	0.25	2	–	–	–
Lift vessel	Jackup	4	2	8	4	2	8
	Pump	–	–	–	4	2	6
	Cut	–	–	–	16 h/m	10 h/m	24 h/m
	Lift	4	2	8	3	2	8
	Move	4	2	8	4	2	8

12.3.4 Parameterization

Vessel dayrates from Table 12.1 are applied. If jackups are used, the spread consists of two tugs and one barge (*SDR* = $21,750/day); if SPIVs are used, the spread consists of one tug and one barge (*SDR* = $11,750/day).[5] OSV dayrates are assumed to range from $5,000 to $10,000 per day, with an average of $7,500/day, including fuel costs. Temporal parameters for both the OSV support and single-vessel models are given in Table 12.6. All parameters are assumed to be constant except for cutting time which varies depending on the monopile diameter. All times are in hours except for CUT time which is in hours per m of monopile diameter [2]. Pump time varies depending on pump capacity; it is assumed that 100–150 m^3 of mud will be pumped at a rate of 25–50 m^3/h. Foundation cuts are assumed to occur 15 ft below mudline. Most parameters are identical regardless of support type, however, lift time proceeds slightly faster if a single vessel is used because the crane will be attached to the foundation concomitant with cutting operations.

12.3.5 Example

Table 12.7 shows the cost range for the removal of a single 5.1 m diameter foundation proposed for Cape Wind. The temporal parameters from Table 12.6 are allowed to vary, but dayrates for the main vessel, vessel spread and OSV are set at their expected values to constrain the output range. It is assumed that a jackup

[5] This represents a reduction in spread requirements relative to turbine removal or installation due to the less sensitive nature of the work and an assumed ability of the main vessel to store at least one removed foundation while waiting for the barge to return from port.

Table 12.7 Foundation removal cost estimates at Cape Wind (1,000 $)

Model	Expected	Minimum	Maximum
OSV support (per foundation)	70.0	38.2	164.9
Single-vessel (per foundation)	345.9	211.1	545.5
OSV support (total farm)	9,100	4,966	21,437
Single-vessel (total farm)	44,967	27,443	70,915

barge is used with the associated barge spread (Table 12.2). The OSV support model cost is approximately 37% of the single-vessel model. Total estimated foundation removal costs for Cape Wind are also shown. Total costs range dramatically with duration assumptions and support methods, but are generally less expensive than turbine removal costs. Foundation removal costs are approximately equal to foundation installation costs if the single-vessel method is used, but are approximately 20% of installation costs if OSV support is employed.

12.4 Cable

12.4.1 Input

The user specifies the inner-array and export cable length, if known; otherwise, distance to shore may be used as a proxy for export cable and inner-array length may be approximated via formula (See Eq. (9.19)). The model provides dayrates, vessel capabilities, spread requirements and activity durations. Historic dayrates and current technologies are assumed. Activity durations are hypothesized relative to empirically derived installation times.

12.4.2 Model

Cable removal cost estimates are identical for export and inner-array cables and is based on the length of cable (in km) divided by the rate of cable removal (in km/day), times the total vessel daily cost.

12.4.3 Parameterization

Cable removal time is estimated from known installation rates (km/day) and an inflation factor (IF); IF = 1 indicates that installation and removal occur at the same rate; IF = 2 indicates that removal proceeds twice as fast as installation. For

Table 12.8 Cable removal cost estimates at Cape Wind

Parameter (unit)	Calculation	Inner-array	Export
Length (km)	Input	130	25
Installation time (km/day)	Input	0.3	0.7
Inflation factor	Input	2	1.25
Removal rate (km/day)	Installation time * Inflation factor	0.6	0.9
Removal time (day)	Length ÷ Removal rate	217	28
Total dayrate ($/day)	Input	32,000	44,000
Expected cost (1,000 $)	Removal time * Total dayrate	6,944	1,232
Maximum cost (1,000 $)	Removal time * Total dayrate	9,244	1,571
Minimum cost (1,000 $)	Removal time * Total dayrate	4,622	783

inner-array cables, IF is assumed to range from 1.5 to 3, and for export cables, IF ranges from one to two. Installation rates generally range from 0.1 to 0.6 km/d (0.3 km/d average) for inner-array cables and from 0.2 to 1.4 km/d (0.7 km/d on average) for export cables [4].

For inner-array cables, the total daily cost is composed of barge spread and OSV cost; diving operations are expected to be supported by the OSV. OSVs are assumed to cost $15,000–$25,000 per day (average $20,000/day) and a barge spread is assumed to cost $7,000–$17,000 per day (average $12,000/day). For export cables, an extra vessel is required to help retrieve buried sections of cable. A variety of vessels could complete this task (e.g. OSV, specialized barge spread, dredge spread) and an additional $15,000/day is assumed to be required.

12.4.4 Example

Table 12.8 summarizes estimated cable removal costs at Cape Wind. Cable length is taken from company estimates [3, 4]. The maximum and minimum costs in Table 12.8 are generated by varying the inflation factor between its maximum and minimum values and leaving all other parameters at their expected values.

12.5 Substation and Met Tower

For substation removal, an OSV will be required to support divers and pile cutting operations and it is expected that the OSV and heavy-lift vessel will remain on site throughout the operation. Therefore, the daily cost consists of the heavy-lift vessel, OSV, and barge spread costs, at a total cost between $122,000 and $140,000 per day. Table 12.9 shows the estimated time of the operation.

Table 12.9 Temporal components of substation and met-tower removal

Activity	Substation (h)	Met tower (h)
Transit	12	12
Place anchor	1	1
Cut topsides	12	4
Lift topsides	3	3
Cut piles	48	36
Lift jacket	3	3
Pull anchor	1	1
Transit	12	12
Total time	92	72
Expected cost*	502	185
Maximum cost*	536	228
Minimum cost*	467	141

Note * Cost are described in 1000 $

For met-tower removal, the topsides are disassembled and placed on a barge and the piles are cut with support from crew on the liftboat. The daily cost is assumed to range from $47,000 to $76,000. The daily cost includes $35,000–$64,000 for a liftboat or jackup and $12,000 for a barge spread. The temporal components are shown in Table 12.9. Substation and met-tower removal cost are estimated to range between $467,000 and $536,000 and $141,000–$185,000, respectively, per unit.

12.6 Scour Protection

Scour protection removal may occur before foundation removal (if an external cut is required) or following foundation removal. A mechanical dredge is required to lift rock armor onto a cargo barge, and for concrete mattresses, dredging methods and costs are likely to be different. Dredging costs for all U.S. Army Corps of Engineer funded projects are approximately $5 per cubic yard (including material disposal). However, nonhopper and work similar to scour protection removal can be more expensive, averaging $23 per cubic yard, with some individual projects costing up to $104 per cubic yard [5]. Given the density of scour protection and the complications of the marine environment, a removal cost of $25 per cubic yard ($\$33/m^3$) is assumed. Each turbine is assumed to be protected with 1,000–1,500 tons of rock armor. Metamorphic rock has a density of approximately 2.75 g/cm^3 [6]; therefore, each foundation is protected by 363–545 m^3 of rock.[6]

[6] This is consistent with the 380 m^3 of scour placed around the turbines at the Horns Rev 2 wind farm.

Given these assumptions, removal cost is expected to range between $12,000 and $18,000 per foundation.

12.7 Site Clearance

Site clearance can be conducted across the entire farm or on a per-turbine basis. That is, the entire area of the farm could be required to be cleared and verified, or just the area around each turbine to some set radius. Current regulations are defined in terms of a structure basis but the radius of clearance has not been specified. Both cases are considered.

12.7.1 Per-Turbine

If site clearance operations for wind turbines are similar to caisson and well-protector site clearance in the oil and gas industry and have a similar level of debris removal,[7] then oil and gas industry data can be used to infer site clearance cost in offshore wind. It is assumed that the area to a radius of 183 m centered at each turbine will be trawled. Since turbine spacing is typically in the order of seven turbine diameters, the clearance zones would not overlap. Site clearance is assumed to be $16,000 per foundation and range from $6,000 to $26,000 [7]. Total clearance cost is determined from the number of turbines, substations, and met towers times the cost per turbine.

12.7.2 Whole Farm

For the entire wind farm, the total project cost is the farm area times the cost per unit area. The area of the wind farm can be input by the user if known or may be estimated via formula.[8] Site clearance and verification cost is assumed to be 84,000 $\$/km^2$ and range from 37,000 to 131,000 $\$/km^2$ [7]. Some savings due to economies of scale are expected.

---

[7] This is believed to be a reasonable assumption because maintenance and repair operations will mostly occur inside the tower and nacelle assemblies and offshore turbines are unmanned facilities similar to caissons and well protectors. The lifetime of the units are also roughly comparable.

[8] Farm area may be approximated by: $AREA = -51.5 + 0.41RD + 0.65NT$, where AREA is described in km^2, RD is the rotor diameter in meters, and NT is the number of turbines.

Table 12.10 Estimated site clearance and verification costs at Cape Wind

Model	Parameterization	Unit cost	Number units	Estimated costs (1,000 $)
Per turbine	Expected	16,000 $/turbine	132[a]	2,112
Per turbine	Minimum	6,000 $/turbine	132[a]	792
Per turbine	Maximum	26,000 $/turbine	132[a]	3,432
Whole site	Expected	43,000 $/km^2	62 km^2	2,399[b]
Whole site	Minimum	19,000 $/km^2	62 km^2	1,060[b]
Whole site	Maximum	67,000 $/km^2	62 km^2	3,739[b]

Notes [a] Includes one met tower and one substation
[b] Includes 10% economy of scale discount

Table 12.11 Disposal options by component

Component	Reef	Landfill	Scrap	In place
Turbine blades	No	Yes	No	No
Turbine nacelle	No	Yes	Yes	No
Turbine tower	No	Unlikely	Yes	No
Monopile-transition piece	Yes	Yes	Yes	No
Monopile	Yes	Unlikely	Yes	No
Cables	No	Yes	No	Yes
Scour protection	No	Unlikely	Yes	Yes
Substation foundation	Yes	Unlikely	Yes	No
Substation topsides	No	Yes	No	No

12.7.3 Example

Application of the site clearance and verification cost algorithm is illustrated for Cape Wind in Table 12.10. The costs of per-turbine clearance are less than the costs of whole farm clearance because the clearance area is smaller, about 23% of the total farm area. Costs range from approximately $1–$4 million with an expected value of $2 million. These costs may overestimate actual cost due to cheaper dayrates associated with the long-term charter of a vessel. Site clearance costs are expected to be a small but non-negligible proportion of removal costs.

12.8 Material Disposal Costs

Disposal costs consist of three components: processing costs, transport costs and scrap profits or landfill costs. There are several disposal options for wind farm components (Table 12.11) and the ultimate disposition of each component will determine total cost.

Table 12.12 Estimated onshore component cutting costs (1000 $)

Component	Thickness (cm)	Cut rate (cm/s)	Material	Labor	Total
Tower	3	0.5	3	0.4	3.4
Monopile	5	0.2	8	1.1	9.1
Monopile-tp*	5 + 7 + 5	0.2, 0.15, 0.2	8 + 10 +8	3.6	29.6

Note * tp denotes transition piece assembly composed of steel, grout, steel
Source [15, 16]

12.8.1 Processing Costs

Material to be scrapped or landfilled incurs processing costs. For landfilled material, processing consists of cutting the pieces into lengths and weights transportable by truck (assumed to be under 40 ft long and under 20 ton). For scrap steel, processing will involve cutting pieces into suitable sizes to be accepted by mini-mills (typically 2 × 5 ft). Processing costs may be estimated on either a per-foot or per-ton basis.

A variety of methods may be used to cut monopiles, towers, and the monopile/transition piece assembly. The cost to cut tubular steel is determined by local labor expense, the complexity of the cut, steel thickness, and the cutting method employed [8]. A variety of tools may be used to make cuts including oxyfuel torches, thermal lances, abrasive water jets, and plasma arcs. Due to the material thickness and assumed lack of mechanized cutting facilities, thermal lances are assumed.

Table 12.12 shows cut time and costs for monopiles, towers and monopile/transition piece assembly. A labor cost of 25 $/h is assumed. The total length of cut required to decompose a tubular member into pieces that can be accepted by a mini-mill is approximated by the formula:

$$\min\left\{\left(\left\lceil\frac{D}{Y}\right\rceil+1\right)L+\left\lfloor\frac{L}{X}\right\rfloor C,\ \left(\left\lceil\frac{D}{X}\right\rceil+1\right)L+\left\lfloor\frac{L}{Y}\right\rfloor C\right\} \tag{12.20}$$

where D is the diameter, C is the circumference, and L is the length of the tubular member, and X the length and Y the width dimension of a final cut piece, all in the same units. The operators $\lceil X\rceil$ and $\lfloor Y\rfloor$ indicate round up and round down, respectively; min indicates the component with the smaller value should be selected. Domestic heavy metal steel HMS #1 accepted by mini-mills is at least ¼ inch thick and no larger than 5 × 2 ft, and so in this case the values of X and Y are 5 and 2 ft (or vice versa). Lengths and diameters may be input or estimated via algorithm.

The cost of removing the grout from the monopile/transition piece is assumed to be $20/ton. For processing of substation jackets a cost of 50–100 $/ton is assumed [8]. Turbine and substation topside processing are more complex and is expected to cost between 100 and 200 $/ton.

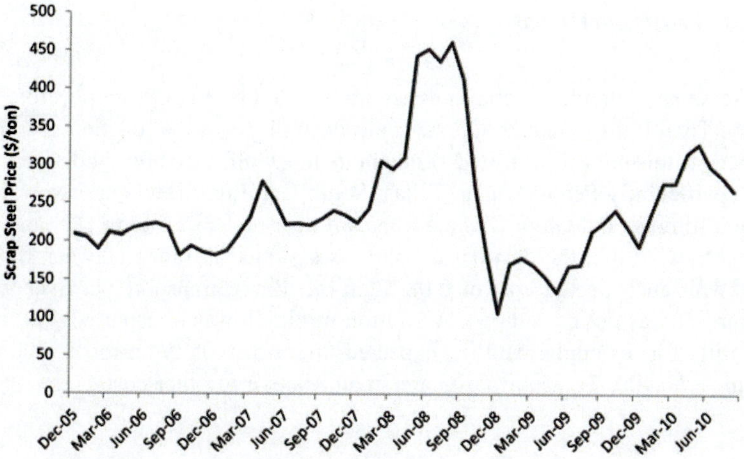

Fig. 12.2 U.S. heavy metal steel scrap prices

12.8.2 Scrap Value

Steel is a highly recyclable material that can be sold at resale determined by the supply and demand conditions in regional scrap steel markets. The value of scrap steel is determined by its weight, quality, size, and the market conditions at the time of sale. The average U.S. price of heavy metal scrap HMS #1 steel from 2005 to 2010 are shown in Fig. 12.2. The average price over the period was 243 $/ton. An average scrap value of 243 $/ton with a range of 100–380 $/ton is assumed in the cost estimation. Copper in cables and turbines have the potential to generate revenue for the decommissioning operation. However, given uncertainties in the methods and costs of removing copper from cables, this is not investigated.

12.8.3 Landfill Cost

Landfill costs are defined as the costs to dispose of anything that is not recycled, reefed, or left in place. Landfilled materials are assumed to fit into the category of "construction and demolition" or inert waste. While some hazardous material such as oil and electronic components exist in turbines, they are not considered. Landfill costs for inert and demolition waste material are variable, and it is assumed that they range from 20 to 136 $/ton with an expected cost of 30 $/ton [9–12]. Landfill costs vary regionally and are expected to be lowest in the Southeast and Gulf and highest in the Northeast and Mid-Atlantic States.

12.8.4 Transport Costs

Both scrap and landfill material must be transported by truck from port to a waste facility. Total transport costs and the transport unit cost are a function of transport distance. Mini-mills are located throughout the Gulf, Atlantic, and Great Lake states, particularly Pennsylvania, Texas, South Carolina, New York, New Jersey, Indiana, Illinois, and Ohio. Given a transport cost of 1.6–2.4 $/mi [12] and that a truck carries 24 ton, the cost per ton-mile is assumed to range between 0.07 and 0.1 $, with an expected cost of 0.08 $/ton-mi. Barge transport via river will be cheaper but was not considered. A 200 ton turbine tower transported 400 mi to a mini-mill, for example, and is expected to cost 0.08 $/ton-mi * 200 ton * 400 mi = $6,400. Transport costs are small relative to other costs.

12.8.5 Reefing

If a reef site can be identified and permitted close to the wind farm, it is possible that reefing will be performed for its ecological value and to reduce disposal, processing, and transport costs. If a reef site is further from the wind farm than the distance to port, additional transport cost will be incurred and reefing is unlikely to be economic. Since neither the distance to port nor the distance to a hypothetical artificial reef is known, it is assumed that the distances are equal and transport costs are ignored. The total costs of reef disposal, including processing, transport, and landfill costs are assumed to be zero.

12.8.6 Example

Table 12.13 shows the expected disposal costs at Cape Wind. It is assumed that the free monopile, monopile/transition piece assembly, and tower are 35, 30, and 225 ft long, respectively, and that the transport distance is 100 mi. The length and weight algorithms derived in Chap. 11 are employed. Due to the low cutting costs and large weight, the tower provides the greatest revenue. However, even with higher processing costs, the monopile/transition piece assembly can generate revenue, suggesting that given this combination of parameters, reefing will not be undertaken. The non-recovered portion of the turbine provides the greatest costs suggesting that economies may be achieved by maximizing the proportion of the turbine recovered. The scrap value of components at Cape Wind after cutting, processing and disposal are expected to achieve a net revenue of $5 million. This value will vary at the time of the operation depending on the steel price and labor costs, and unlike the other elements of decommissioning, have the potential to generate revenue.

Table 12.13 Wind farm components, disposal weights, and expected cost and revenue at Cape Wind

Component	Weight (ton/unit)	Total farm weight (ton)	Cut length (ft/unit)	Disposal method	Cutting costs ($/ft)	Processing costs ($/ton)	Transport costs ($/ton-mi)	Landfill costs ($/ton)	Scrap profit ($/ton)	Total revenue (million $)
Monopile	71	9,234	717	Scrap	9.1	–	0.08	–	243	1.3
Monopile/transition	107	12,766	615	Scrap	29.6	–	0.08	–	243	0.7
Grout	5	650	–	Landfill	–	20	0.08	30	–	−0.04
Tower	190	24,700	4,108	Scrap	3.4	–	0.08	–	243	4.0
Turbine[a]	132	17,160	–	Scrap	–	150	0.08	–	243	1.4
Turbine[b]	88	11,440	–	Landfill	–	150	0.08	30	–	−2.1
Export cable	75 kg/m	2066	–	Landfill	–	–	0.08	30	–	−0.08
Inner-array cable	30 kg/m	5026	–	Landfill	–	–	0.08	30	–	−0.19
Substation jacket	900	900	–	Scrap	–	75	0.08	–	243	0.14
Substation topsides	1000	1000	–	Landfill	–	150	0.08	30	–	−0.15
Total revenue										5.0

Notes [a] 60% scrapped
[b] 40% landfilled

Table 12.14 Expected cost for three decommissioning scenarios at Cape Wind

Stage	Component	With cable removal (million $)		Without cable removal (million $)	
		Cost	Revenue	Cost	Revenue
Removal	Turbines	48.7 (22.1)	–	48.7 (22.1)	–
	Foundations	9.1	–	9.1	–
	Substation	0.5	–	0.6	–
	Cable-export	1.2	–	–	–
	Cable-inner array	6.9	–	–	–
	Met tower	0.2	–	0.3	–
	Scour	1.9	–	–	–
	Site clearance	2.1	–	2.1	–
Removal subtotal		70.5 (43.9)	–	60.6 (34.0)	–
Disposal	Turbine–nacelle and blades	2.1	1.4	2.1	1.4
	Turbine-tower	–	4.0	–	1.4
	Foundation	0.04	2.0	0.04	4.0
	Substation jacket	–	0.14	–	2.0
	Substation topsides	0.15	–	0.15	0.14
	Cable–export	0.08	–	–	–
	Cable-inner-array	0.19	–	–	–
Disposal subtotal		2.5	7.5	2.3	7.5
Total cost		65.5 (38.9)		55.4 (28.8)	

Note Felling option cost denoted in parentheses

12.9 Cape Wind Decommissioning Scenarios

Table 12.14 shows three scenarios for decommissioning operations at Cape Wind. In the first scenario, the cable and scour protection is removed; in the second scenario, cable and scour protection is left in place; in the third scenario, the turbine felling option is employed. The felling option cost is denoted in parentheses.

The first two scenarios are parameterized with expected values. In both cases the costs are similar and the majority of costs are due to turbine removal. It is assumed that foundation cutting will be performed by an OSV; if a heavy-lift vessel is required, the total decommissioning costs increase by approximately $35 million and foundation removal becomes as expensive as turbine removal. In both scenarios, it is the removal costs rather than the disposal costs that drive overall decommissioning costs. Overall costs are approximately 100,000–140,000 $/MW; these costs are less than 5% of the total capital costs of the development (estimated at 3.8 million $/MW).

Removal costs range from $49.8 to $100.1 million with an expected value of $70.5 million, assuming cables are removed and standard removal methods are used. If cables are left in place, removal costs range from $44.4 to $89.3 million with an expected cost of $60.6 million. Disposal costs with cable removal range from $7.9 million to a $15.7 million profit; without cable removal costs range from $6.9 million

Table 12.15 Expected decommissioning costs at proposed U.S. wind farms

Wind farm	Capacity (MW)	Number turbines	Distance to port (nm)	Removal costs (million $)	Disposal costs (million $)	Scrap revenue (million $)	Total cost (million $)
Coastal Point	150	60	20	24.5	0.9	1.9	23.4
Bluewater	450	150	100	68.3	1.7	10.3	59.7
Garden State	350	96	80	47.7	1.8	4.2	45.3

to a $15.9 million profit. Total decommissioning costs range from $34.1 to $108 million with cable removal and $28.5 to $96.1 million without cable removal.

The results are similar to Cape Wind's estimate of decommissioning costs, despite quite different methodologies. Cape Wind used a generic estimate of decommissioning cost of £40,000 per MW, taken from a European study based on a series of interviews with industry personnel [13]. Cape Wind doubled the estimated unit cost and, after accounting for inflation and the exchange rate, settled on a gross cost of $66 million (141,000 $/MW). Cape Wind estimated the scrap value of the farm as $37 million for a net cost of $29 million [14].

Cape Wind's estimated removal costs of $66 million are similar to the model estimate of $70.5 million. However, the Cape Wind developers assumed the scrap value of project components to be $37 million while the present study estimated a net scrap revenue of $5 million. If copper from the cables is recovered the scrap revenue estimates presented here may be low [13, 14], but not to the extent accounted by this difference.

The turbine felling option is also depicted. Under felling, the turbine removal costs decline from $48.7 to $22.1 million. The range of turbine removal costs using the felling technique is $8.8–$35.1 million; even in the maximum parameterization felling is less expensive than the expected costs via standard methods. This suggests that the felling option will be a less expensive alternative to standard decommissioning methods if the approach is feasible and technical obstacles can be overcome.

12.10 Decommissioning Costs at Proposed Offshore Wind Farms

Table 12.15 shows decommissioning costs at three proposed U.S. offshore wind farms: Coastal Point (TX), Bluewater Wind (DE), and Garden State (NJ). Input data on the expected system configurations are obtained from public sources, but because of the early stages of development, the input is significantly more uncertain than Cape Wind, and thus, the output results are similarly limited.

Turbines are assumed to be removed by SPIV via the self-transport method. Foundations are removed by OSV support, cables and scour are left in place, and expected values are used for all parameterizations. Two of the projects plan to use jacket or tripod foundations rather than monopiles; removal costs for these projects are modeled as if they were monopiles.

Estimated decommissioning costs at the three projects ranged from 129,000 to 156,000 $/MW, consistent with estimates for Cape Wind. As in Cape Wind, removal costs are the dominant driver of total cost, with turbine removal accounting for approximately 80% of removal costs. The model output is insensitive to changes in weight, length and diameter assumptions, but is sensitive to changes in assumptions that impact turbine removal time. When the development plans for these offshore wind farms are better established, it will be worthwhile to revisit the cost estimates using better input data.

12.11 Discussion

Bonding requirements for offshore wind farms are most accurately estimated on a case-by-case basis using models like those presented here. It would be inappropriate to determine bonding requirements on a per-MW or per-turbine basis as decommissioning costs are dependent on the turbine size and capacity, the number of turbines, foundation type, methods of removal, distance to port, methods of disposal, and the components allowed to remain in place.

From a regulatory perspective, it is important that operators remain a going concern and generate enough income to satisfy decommissioning requirements at the end of the lease. In the offshore oil and gas industry, an important measure of the capability of an operator to meet their decommissioning obligations at any point in time is the ratio of reserves or annual income to their decommissioning liability. In offshore wind, the equivalent measures would be the value of power expected to be generated. Power Purchase Agreements (PPA) are finalized before construction begins, and capacity factors can be reliably estimated from wind speed data. Therefore, unlike oil and gas development, the income of the developer may be reliably forecast well into the future which should serve to reduce uncertainty associated with default of the operator.

Decommissioning cost models are limited by the need to project current costs and technologies into the future. While prices may be adjusted for an expected inflation rate, vessel dayrates are highly uncertain and have significant direct and indirect impacts on costs. Due to these uncertainties, the range of all model parameters is large. As uncertainties propagate through the calculations, the range of model outputs increases. Removal technologies are not expected to change dramatically over the time periods considered, however, and so under a traditional scenario changes due to technological improvements are considered to be small. Novel and creative methods of turbine removal, however, may reduce total decommissioning cost, but progress along this front is not likely to be made until significant decommissioning activity starts to occur.

References

1. Hewson TA, Stamberg JB (2008) Beech Ridge Energy LLC financial assurance needs. Prepared for Mountain Communities for Responsible Energy. 7 October.
2. Kaiser MJ, Pulsipher AG, Byrd RC (2004) Model estimates abrasive cutting costs. Oil Gas J 102(39):44–47
3. MMS (2008) Cape Wind final environmental impact statement. Minerals Management Service, Herndon, VA. 2008-040
4. Pirelli ABB (2004) Interstitial optical option for submarine composite cable. http://www.boemre.gov/offshore/PDFs/CWFiles/123.pdf. Accessed 24 Oct 2010
5. USACE (2010) Analysis of dredging costs. U.S. Army Corps of Engineers. http://www.ndc.iwr.usace.army.mil//dredge/ddcosts.htm. Accessed 24 Oct 2010
6. Smithson S (1971) Densities of metamorphic rocks. Geophysics 36:690
7. Kaiser MJ, Martin J (2009) An update on the cost of net trawling operations in the Gulf of Mexico. Marine Policy 33(1):58–63
8. Kaiser MJ, Pulsipher AG (2010) Scrap and storage markets in the Gulf of Mexico-III. Process workflows and industry structure. Energy Sources, Part B: Econ, Plan Policy 5(3):250–260
9. Araman PA, Bush RJ, Hammet AL, Hager EG (1998) Wood pallets and landfills: Status and opportunities for economic recovery and recycling. In: Proceedings, WASTECON/ISWA World Congress, SWANA 36th Annual International Solid Waste Exposition, Charlotte, NC, Solid Waste Association of North America. Silver Spring, MD, 26–29 Oct 1998
10. Elzarka H (2007) Making the case for construction waste management. In: International Proceedings of the 43rd Annual Conference of the Associated Schools of Construction. Flagstaff, AZ
11. Georgia Department of Community Affairs (2010) Annual solid waste report—executive summary. http://www.dca.state.ga.us/development/research/programs/downloads/2009ExecutiveSummary.pdf. Accessed 24 Oct 2010
12. Fairtran (2010) Rates and indexes. https://www.fairtran.com/fairtrans/Truckload-Rate-Index.aspx. Accessed 24 Oct 2010
13. Climate Change Capital (2006) Offshore renewable energy decommissioning study. Climate Change Capital p 75
14. Rosenzweig DS (2010) Affidavit of Dennis J. Duffy to Massachusetts Department of Public Utilities. Docket 10–54, August 31
15. Wang H, Pranda P, Hlavacek P (2004) Flame-focusing modification of a wire-core thermal lance. Weld Res 39(3):283–287.
16. Wang H, Pranda P, Hlavacek P (2004) The eutectic generation effect and chemical modification of thermal lance cutting of concrete. J Mater Sci 39(23):7003–7008